A Guide to Career Resilience

Eve Sprunt · Maria Angela Capello

A Guide to Career Resilience

For Women and Under-Represented Groups

 Springer

Eve Sprunt
Dublin, CA, USA

Maria Angela Capello
Houston, TX, USA

ISBN 978-3-031-05590-4   ISBN 978-3-031-05588-1   (eBook)
https://doi.org/10.1007/978-3-031-05588-1

This Springer imprint is published by the registered company Springer Nature Switzerland AG
The registered company address is: Gewerbestrasse 11, 6330 Cham, Switzerland

For my children
Alexander Dalziel Sprunt and Elsa Dunbar Sprunt Broeker

Eve Sprunt

For my beloved husband, Herminio Passalacqua,
father of the jewels in our life together: Alessandra and Claudia

Maria Angela Capello

Foreword

I am honored and privileged to write this foreword for the new book by Eve Sprunt and Maria Angela Capello, because they are addressing a very important topic on building resilience in navigating one's career. I believe that everyone needs a mentor, coach, and a promoter or sponsor to build a successful career. However, as is well illustrated in a very authentic way in this book, not all mentoring or sponsoring has positive effects. Without careful attention, these types of relationships can lead to devastating outcomes. **A Guide to Career Resilience for Women and Under-represented Groups** offers essential advice for what women should do when they find themselves in a mentoring and/or sponsoring relationship that has gone sour, with practical insights, reflections, and tips that are applicable to women as well as men. The authors provide examples of common challenges people face in their careers and offer advice on how to get people around you to advocate for you, including tips on how to get out of or get through a rut.

Yet, knowledge is power! I wish I had had a book like this early in my career. After serving in many leadership roles at multiple universities and professional organizations, I know progress in our careers is a product of not only our hard work which leads to professional accomplishments but also the mentoring and career guidance we receive through our networks. Good mentoring and effective sponsoring can make a tremendous difference in life, especially for women. Even now, when we believe we have made significant progress toward parity, many of the challenges faced in the past remain as pulsating issues today. That is why I believe this book is extremely timely and a must read! To be successful, we must leverage all the advice and support we can get and avoid getting trapped in toxic environments.

Eve and Maria Angela have done a thorough job of including real experiences from those who have suffered from bad mentoring or sponsoring. Some of their interviewees were still suffering, distressed, and traumatized, years after a disastrous relationship. They have also shared their own powerful lived experiences, which they have not previously divulged. For this book, they plunged deep into their painful memories related to mentoring to share the valuable, but difficult, lessons they learned. Those lessons forged who they are today and stoked their shared passion for helping others navigate similar situations. In so doing, they have achieved something truly remarkable: a compilation of real, unadulterated stories that have the power to help others avoid the same pitfalls. Their goal is to enable others to recognize toxic situations so that they can escape before permanent damage occurs. Having mentors and sponsors is essential to success, but we must be prepared to escape if the relationship goes bad.

Reading this book prompted self-reflection, because every chapter describes in very personal ways situations they have encountered and escaped. One of my favorite parts is the one about "queen bees," because learning early in your career to differentiate sincere and supportive mentoring from superficial posturing is essential to avoid damage to your reputation, self-esteem, and mental health.

It is easy to discuss the benefits of mentoring and sponsoring, which are many, but not so simple to share the dark side of these loops. I offer my personal thanks to Eve and Maria Angela for sharing what they have learned about both the bright and somber sides of mentoring and sponsoring and how to avoid the pitfalls they and their anonymous interviewees encountered. They have shared comprehensively and courageously, providing readers with tools that show them how to navigate virtually any career valley and come out the other side while also providing guidance on how to build a successful career.

I admire Eve and Maria Angela's resilience and appreciate their willingness to honestly share some of the darkest moments of their lives so that readers may benefit. My experience with both is that they strive to support others with initiatives like this unusual and insightful book. In the end, this book celebrates the positive mentors and sponsors but is also a must read for the less-than-ideal ones. The hope is that by understanding the lived experiences of those that have received negative mentoring, they also can use this book to unlearn negative attitudes which can help them provide more positive mentoring to future mentees.

Estella Atekwana
Dean, College of Letters
and Science
University of California
Davis, USA

Preface

When we wrapped up the first book we co-authored together, **Mentoring and Sponsoring, Keys to Success**, we knew there was a lot more we wished to explain about these critical processes. People love to share their successes, but there is much to be learned from what can go wrong and how people have survived unpleasant situations and come back stronger. Several of the people we wanted to interview for our first book politely declined for undetermined reasons or simply decided not to tell us why. We realized they were not comfortable revealing their own experiences with mentors and sponsors. This motivated us to focus on issues about mentors or sponsors that can adversely affect people. We have faced challenging situations, and we realized other very successful people who, even now, were reluctant to disclose their story have also struggled with difficult advisors and supporters.

Is silence the answer? We soon realized it was not and felt the need to share our own struggles and those of our anonymous sources who trusted us with their painful stories, so that you can avoid some pitfalls that plagued all of us. We wanted to complement the stories of our first book with this additional information that describes how we survived (yes, survived!) and overcame some extremely stressful and painful situations. If you encounter problems, we hope the stories of resilience that we include here will inspire you to break out of your depression and push on to achieve your goals.

Throughout this book, we provide our personal insights and experiences as part of the narrative in the plural form of the first person. We use "we" and "our" in every chapter because we experienced and surmounted many of the challenges we describe. We believe that describing our personal struggles along with those of our interviewees, who graciously spoke with us about some of their darkest moments but wish to remain anonymous, will provide you with valuable guidance so that you can thrive in any challenging work environments you encounter.

We hope our book will benefit a wide range of people, especially women and members of underrepresented groups who, despite tremendous advances toward equality, are more likely to encounter special challenges interacting with mentors and sponsors.

Dublin, USA   Eve Sprunt
Houston, USA   Maria Angela Capello

Acknowledgments

Eve and Maria Angela would like to thank the women and men, who shared their unpleasant mentoring and sponsoring experiences. They will remain anonymous, but we most sincerely appreciate their courage in trusting us with their stories of some of the most painful episodes in their lives.

Also, we want to thank Hugh, Eve's husband, and Herminio, Maria Angela's husband. They were very understanding of the time our writing required and were patient with being ignored during normal working hours as well as evenings and weekends.

Dr. Alexis Vizcaino was the first supporter of this initiative. Even after writing the thousands of words in this book, we remain wordless as to how to express our appreciation for his decisive endorsement. Thank you!

When we discussed potential images for the cover, we decided that a picture of a queen bee would be perfect. During our careers, Eve, Maria Angela, and many of the people we interviewed have suffered the stings of vicious human queen bees. Despite the queen bees' attempts to destroy our self-confidence and our careers, their brutal attacks made us stronger and more compassionate. Instead of destroying our self-confidence and ability to succeed, they made us more resilient. For this, we realized we also owe them our thanks.

Margarita Alberdi de Genolet and her husband, Luis Carlos Genolet, graciously provided the photograph for the cover of our book.

We would also like to thank Dr. Estella Atekwana for graciously providing a Foreword for our work. She is a valued friend in our network.

Susan Howes is one of the most effective networkers we know. She has provided valuable feedback and suggestions on the manuscript for this book and on the manuscripts for all of Maria Angela's previous books including the one I co-authored with her.

These acknowledgments are a preview of how important we think a strong network is for success in almost any endeavor you undertake. We treasure the members of our networks.

The Photos of Our Book

A honeybee colony is an amazing example of networking, mentoring, and busy workers in a structured organization. Images of beehives are mesmerizing with myriad bees flying about industriously. Thousands of worker bees cooperate in building the hive, food collection, and brood rearing.

We selected photos of bees to illustrate our chapters with the aim of prompting reflections about the individual roles of the working bees, the drones, and especially, the queen bee, given its human counterpart's role in disappointing sponsoring and mentoring relationships.

The cover photo shows a Queen Bee surrounded by her court in the apiary of Margarita Alberdi de Genolet, a former geochemist who re-invented herself and became a renowned apiculturist and now lives happily among hives and bees. Every year, the queen bees in her apiary in Kassel, Germany, are marked with color dots, to facilitate their differentiation. In 2021, the designated color was white. A hive needs a queen to survive. The female worker bees that surround the queen are called the "court," and they take care of cleaning and feeding the Queen. Margarita's husband, Engineer Luis Genolet, took the photograph and graciously agreed to let us use it for this book. He also provided the other photos that we used for the chapters of our book.

Other photos were by Damien Tupinier, Delia Giandeini, and Massimiliano Latella on the site Unsplash, who we thank for sharing their photographic art free of royalties.

We hope you will enjoy the photos we selected to illustrate our book as much as we were delighted learning about the bees and selecting each one of the photos.

The men of experiment are like the ant; they only collect and use. But the bee... gathers its materials from the flowers of the garden and of the field but transforms and digests it by a power of its own.
                    —Leonardo da Vinci

Contents

About the Authors

Eve Sprunt, Ph.D. is a prolific writer and speaker who has inspired many people, challenged huge corporations about gender pay equity, and pushed multiple not-for-profit organizations to be more inclusive. She has extensive scientific and managerial experience and has served as the 2006 President of the Society of Petroleum Engineers, for a year as President of the American Geosciences Institute (2017–2018), was the founder of the Society of Core Analysts, and served as the Vice-President of the Society of Exploration Geophysicists. She received the top technical award of the Society of Women Engineers in 2013, the Achievement Award, and in 2010 the Society of Petroleum Engineers' top award, Honorary Membership. The Association for Women Geoscientists honored her in 2015 as the recipient of the ENHANCE Award "for life-long commitment to advancing the status of women in the geosciences."

She received her bachelors and master's degrees from the Massachusetts Institute of Technology (MIT) and in 1977 was the first woman to receive a Ph.D. in Geophysics from Stanford University. She has authored more than 120 editorial columns on industry trends, technology and workforce issues, 23 patents, and 28 technical publications.

During her career in the energy industry with Mobil and Chevron, she worked in a wide range of areas, including university recruiting and philanthropy, venture capital investing, alternative energy, climate change policy, business development, technical service and research. After retiring from Chevron, she has focused on helping young professionals by sharing her insights on critical career issues that impact women and dual-career couples. She is the author of **A Guide for Dual-Career Couples, Rewriting the Rules** and **Dearest Audrey, An Unlikely Love Story** (which is a true story about a woman's choice between her career and the love of her life) and co-author with Maria Angela Capello of **Mentoring and Sponsoring: Keys to Success**. A survivor of metastatic neuroendocrine cancer and breast cancer, she enjoys hiking in the hills around her home in the San Francisco Bay Area of California and visiting her grandchildren.

Maria Angela Capello, Cavaliere OSI is a renowned leader and author, expert in Sustainability, Reservoir Management, Diversity and Inclusion, and Digital Transformation. She is a thought leader who advocates for step-changes in sustainability and the effective inclusion of underrepresented groups in the energy sector, particularly women and Latin American professionals. She is the lead author and is advancing the *Geophysical Sustainability Atlas*, a step-change in the visualization of Geoscientists as pivotal agents of change in advancing the UN Sustainable Development Goals.

She is Co-Chair of the United Nations UNECE Committee for Women in Resource Management and is Chair of the Sustainability Committee of the Society of Exploration Geophysicists. She is founding partner of Red Tree Consulting LLC, a boutique consulting firm specialized in sustainability initiatives, and her clients have included oil and gas operators, research centers, and universities.

She is the lead author of two books: **Learned in the Trenches: Insights into Leadership and Resilience** (2018) and **Mentoring and Sponsoring: Keys to Success** (2020) and was the first woman to supervise seismic crews in the jungles of Venezuela, to advance a global career reaching leading roles in Kuwait Oil Company, Halliburton, and PDVSA. She is the only person to have been Distinguished and Honorary Lecturer for the three main professional societies in oil and gas, AAPG, SEG, and SPE.

She is a Cavaliere dell'Ordine della Stella d'Italia (Knight of the Order of the Star of Italy), the highest civil honor conferred by the President of Italy to Italians abroad. She is an Honorary Member of the Society of Petroleum Engineers (SPE), the top individual award of the SPE, and serves on the Board of Directors of the Foundation of Society of Exploration Geophysicists (SEG). She is a Physicist from the Universidad Simon Bolivar (Venezuela), a M.Sc. of the Colorado School of Mines (USA), and is certified in circular economy and sustainability by the University of Cambridge (UK). She is also a pianist, specialized in the baroque period.

Part I

Introduction

Last night as I was sleeping, I dreamt—
O, marvelous error—
That there was a beehive here inside my heart
And the golden bees were making white combs
And sweet honey from all my failures.

—Antonio Machado

1

Multi-decade Perspective

Life is a marathon, not a sprint. Over a multi-decade career, our interests and aspirations evolve and our personal preferences and constraints change. We can begin our careers as strong players, but if we are not resilient and able to adapt to changing circumstances, we are unlikely to be able to achieve long term success. Our goal in this book is to provide advice on how to overcome a wide range of obstacles that you may encounter. This advice is useful to both men and women, but as women, who have spent our careers working in male-dominated organizations, we know what it feels like to belong to an under-represented group. So, we believe our book is especially valuable for members of under-represented groups, particularly for women, because they are often under-represented in positions of authority or leadership.

In 2015, the United Nations launched as Sustainable Development Goal #5 (SDG 5) "Gender Equality" as one of the global goals of humanity for 2030. However, around the world, women still feel the need to create organizations, circles and mentoring initiatives, to address challenges they believe make it more difficult for them to succeed than men. A recent example of one of these groups is Mothers in Science[1] (an international nonprofit organization registered in France which was founded by Isabel Torres and Sonal Bhadane in 2019). Mothers in Science assembled a multinational team in science, technology, engineering, mathematics and medicine (STEMM) including people based in France, US, Canada, Singapore, Portugal, UK,

---

[1] www.mothersinscience.com, as of November 2021.

© The Author(s), under exclusive license to Springer Nature Switzerland AG 2022
E. Sprunt and M. A. Capello, *A Guide to Career Resilience*,
https://doi.org/10.1007/978-3-031-05588-1_1

Nigeria, Finland, Germany, Brazil and New Zealand. Their focus is advocating for mothers in STEMM and creating evidence-based solutions to promote workplace equity and inclusion of parents and caregivers.

Mothers in Science came to our attention because the Association for Women in Science[2] (which was founded in 1971 and is based in the US) hosted in November 2021 a webinar[3] so that Mothers in Science could share the results of their recent surveys. Dr. Torres highlighted how underrepresented women are in STEM. Her conclusion is that the major leak in the pipeline of female talent is when women start a family.[4] Their survey in nine different languages of about 9000 people around the world was conducted via social media and through their network. They found that 34% of mothers stop working full-time when they have children. Dr. Torres said attrition in the United States was worse than average with 42% of mothers ceasing to work full-time. The survey was conducted prior to the COVID-19 pandemic, so they conducted a worldwide poll of mothers and learned over 90% found closure of schools adversely impacted their work productivity. Dr. Torres said, "*Mothers are the default parent when childcare is lacking…and only 26% can rely on their partner to look after the children.*"[5] Furthermore, she found that in every country in which studies have been conducted, "*Mothers have a much bigger share of childcare and housework*" than their heterosexual partners.

While far more women are working in STEMM than when we joined the workforce, it is troubling that the attrition of mothers is still so high. Dr. Torres' observations meshed with Eve's and Maria Angela's. We believe women will benefit from the practical advice in our book on how to be more resilient and overcome the obstacles they will face during their careers. Over the course of a long career, almost everyone will face adversity, but the odds are that women will have to struggle harder than men to succeed.

The McKinsey report Women in the Workplace 2021, which consolidates data from 65,000 employees around the world, concluded that,

> In spite of the challenges of the COVID-19 crisis, women's representation improved across all levels of the corporate pipeline in 2020. This is an encouraging sign—and worth celebrating after an incredibly difficult year. But there are also persistent gaps in the pipeline: promotions at the first step up to manager are not equitable, and women of color lose ground in representation at every level.

---

[2] www.awis.org, as of November 2021.

[3] https://www.awis.org/resource/motherhood-causing-critical-leak-stem-pipeline/, as of November 16, 2021.

[4] Ibid.

[5] Ibid.

The study then emphasizes that there are even more gaps if race is considered, besides gender: "There is still a 'broken rung' at the first step up to manager. Since 2016, we have seen the same trend: women are promoted to manager at far lower rates than men, and this makes it nearly impossible for companies to lay a foundation for sustained progress at more senior levels. Additionally, the gains in representation for women overall haven't translated to gains for women of color. Women of color continue to lose ground at every step in the pipeline—between the entry level and the C-suite, the representation of women of color drops off by more than 75 percent. As a result, women of color account for only 4 percent of C-suite leaders, a number that hasn't moved significantly in the past three years."[6]

Another recent report, published in December 2021 by BCG (Boston Consulting Group) and the World Petroleum Council, focused on women in the oil and gas sector. They found that the percentage of women on Boards in that sector remains low in comparison with other sectors, despite a 22% jump from 2017.[7] In addition, the same report researched best practices for improving diversity and inclusion and listed mentoring programs as the first pivotal element for enhancing retention of women in the workforce.

On almost any day, the news also includes stories about sexual harassment and mistreatment of female employees. The same week as Dr. Torres' webinar, the Wall Street Journal published an article entitled, "Activision CEO Bobby Kotick Knew for Year About Sexual-Misconduct Allegations at Videogame Giant"[8] The article stated, "Activision has been thrown into turmoil in recent months by multiple regulatory investigations into alleged sexual assaults and mistreatment of female employees dating back years... the California Department of Fair Employment and Housing filed a lawsuit in July [2021] alleging that the company ignored numerous complaints by female employees of harassment, discrimination and retaliation, citing what it called its 'frat boy' culture." Activision is not a small company. It is the second largest publicly traded videogame company by market capitalization and employs about 10,000 people.[9]

In August 2021, Activision named a longtime employee, Jennifer Oneal to be the first woman to co-lead one of the company's business units. "Ms. Oneal said in the email [sent in September] she had been sexually harassed

---

[6] https://www.mckinsey.com/featured-insights/diversity-and-inclusion/women-in-the-workplace, as of December 11, 2021.

[7] https://www.bcg.com/publications/2017/energy-environment-people-organization-untapped-reserves, as of December 11, 2021.

[8] https://www.wsj.com/articles/activision-videogames-bobby-kotick-sexual-misconduct-allegations-116 37075680, as of November 16, 2021.

[9] Ibid.

earlier in her career at Activision, and that she was paid less than her male counterpart at the head of Blizzard (a subsidiary) and wanted to discuss her resignation. '*I have been tokenized, marginalized, and discriminated against,*' wrote Ms. Oneal, who is Asian-American and gay."[10]

In other sectors, for example politics, woman face blatant discrimination and/or unconscious bias. In April 2021, a notorious diplomatic protocol incident, "Sofagate,"[11],[12] occurred during the visit of Ursula von der Leyen, President of the European Commission and President of the European Council, with Charles Michel to Turkey. There were only two chairs prepared for the three leaders in the room in which they were received, so as a real time logistical solution for the impasse President of the European Commission, Ms. Ursula von der Leyen, was offered a seat in the same room on the sofa. A series of reactions immediately reached the media and made the video of the incident go viral. This incident highlights that bias against women is still present and a problem at even at the highest levels.

To survive and thrive throughout our careers, we all need to be resilient. We must learn to "roll with the punches" and fight back. We must be prepared to get up out of the dirt and reinvent ourselves.

---

[10] Ibid.

[11] https://www.bbc.com/news/av/world-europe-56696618, as of December 11, 2021.

[12] https://en.wikipedia.org/wiki/Sofagate, as of December 11, 2021.

2

Business Cycles

Careers typically span multiple decades. During that time, the overall economy will gyrate through several business cycles. The time from one economic peak to the next, or one recessive trough to the next, constitutes a business cycle. The National Bureau of Economic Research (NBER) of the United States[1] identified eleven cycles between 1945 to 2009 with the average cycle lasting a little over 5-1/2 years. Some cycles are shorter, and some are much longer. The pandemic caused a recession that ended a relatively long business cycle. In the United States GDP dropped 31.4% in the second quarter of 2021, but the economy rebounded within two months.[2] According to the official government website of the United States Bureau of Economic Analysis,[3] in 2020, the gross GDP of the United States dropped 5.1% in the first quarter and 31.2% in the second quarter but increased 33.8% in the third quarter and continued to improve through the end of 2021. The economic shock in April, May and June was more than three times as sharp as the previous record—a drop of 10% in 1958—and nearly four times the worst quarter during the Great Recession.[4]

---

[1] https://www.nber.org/research/business-cycle-dating.

[2] https://www.cnbc.com/2021/07/19/its-official-the-covid-recession-lasted-just-two-months-the-shortest-in-us-history.html.

[3] https://apps.bea.gov/iTable/iTable.cfm?reqid=19&step=2#reqid=19&step=2&isuri=1&1921=survey, as of January 28, 2022.

[4] GDP Drops At 32.9% Rate, The Worst U.S. Contraction Ever: Coronavirus Updates: NPR.

E. Sprunt and M. A. Capello, *A Guide to Career Resilience,*
https://doi.org/10.1007/978-3-031-05588-1_2

Individual industries may have their own business cycles, which can be countercyclical to the general trend of the economy. Almost all of us will encounter speed bumps and downturns in our careers related to changes and trends that are beyond our control. To be resilient, we must mentally prepare ourselves to face adversity.

If you work in academia, you may think that business cycles will not affect you, but they can. Just as there are business cycles, there are birth cycles with baby booms[5] in which the number of births rise significantly, and baby busts[6] in which the birth rate falls. These changes in birth rate impact the demand for education in the future. Also, the Covid-19 Pandemic has had a huge impact on education.[7] Distance learning technologies introduced during the Pandemic are likely to have long-lasting consequences for educational establishments.

Business cycles have a big impact on job markets. Trends within industries and technical progress also impact the demand for your skills and services. Over the course of a career, you can expect to experience multiple cycles, some of which may have a much larger swing between peak and trough. The longer and deeper the recession and the lower the demand for your skills and experience, the more resilience you will need.

Not surprisingly, in good economic times, when companies are actively trying to recruit employees, people with sought-after skills receive multiple job offers. Hiring incentives and working conditions tend to be much better. People with weaker technical and interpersonal skills can secure employment and retain their positions during periods of high demand for people in their industry. If someone thinks that they are being mistreated, they can easily find another position during buoyant economic periods.

However, when a downturn hits, and jobs become scarce, employers no longer must try as hard to attract and retain talent. Employees who have never experienced a downturn may be shocked by the changes in their work environment. Employers may quickly shift from competing for scarce talent to laying off their staff. When jobs are scarce, employees may feel that they must tolerate treatment that would normally send them racing for the exits. In downturns, morale and working conditions tend to deteriorate.

Good, highly skilled people may find themselves unemployed if their company decides to make massive staff cuts to reduce costs. Often when one employer decides certain job functions must be cut, other similar companies

---

[5] https://en.wikipedia.org/wiki/Baby_boom, as of December 2021.

[6] https://www.newstatesman.com/politics/2021/07/baby-bust-how-declining-birth-rate-will-reshape-world, as of December 15, 2021.

[7] https://tcf.org/content/commentary/global-view-pandemics-effect-higher-education/?agreed=1.

are responding to the downturn in the same way. To survive, you may have to swallow your pride and explore lower paid, less attractive jobs that don't use all your previously highly compensated skills.

During downturns in your profession or the economy overall, demand for your experience and skills may decline. Downsizing measures in private companies are often unavoidable and painful ways to cut costs. Even very strong performers may lose their jobs or be demoted, because their skills are no longer needed. When almost everyone is worried about their retaining their position, morale suffers.

All around the world, if you work for the government, or in a government owned company run by the government of your country or in the health and education sectors, you may expect more stability but lower salaries than in a private company. Consider your risk tolerance when you compare job opportunities in the public versus private sectors.

Even if you are lucky enough to enjoy a sustained period of high demand for your skills, periodically explore new opportunities beyond your current employer to maintain your awareness of other ways in which you might apply your talents. It is always a good idea to frequently take stock of your skills and the demand for them, because acquiring new skills and credentials takes time.

Mentors, who have survived one or more business cycles can be extremely helpful in anticipating changes. An experienced mentor may be able to provide early warning of layoffs because of downsizing and help you identify new opportunities.

2.1   Add to the Bottom Line

During the expansionary, "boom time" phase of a business cycle, an organization can accumulate a lot of "fat". When there is plenty of money, managers may find it easier to ignore troublemakers and under-performing staff. At the beginning of a recession, those people are often the first to go and the impact on the survivors is minimal. The remaining workers are not demoralized, because many of them may have been wondering how the poor performers lasted as long as they did.

The longer and deeper the downturn, the more difficult the staff reductions become. When surviving staff see the termination of people who they consider to be strong contributors, they may ask themselves, "*Why them and not me?*"

Eve survived two rounds of layoffs in which half the people in her group lost their jobs. The remaining staff members including Eve were demoralized and frightened. After the first of those two devastating layoffs, Eve asked to leave research and switch to a business function. She wanted to "get closer to the money-making part of the company." The evening before the second fifty percent layoff, Eve spoke with a colleague who was still working in the job function Eve fled when the first 50% layoff occurred. He was confident that he would not lose his job, because they were short-staffed and he was working overtime. What he didn't appreciate was that the company had decided to outsource that technical function. He was terminated and offered part-time work as a contractor.

When times are tough, organizations focus on their core strengths to survive. A tried-and-true method for individuals to survive is to "follow the money" and work on projects that are critical to the financial success of the organization.

2.2   Be a Problem Solver or Spot a Problem and Solve It!

The famous architect, I. M. Pei said, "*Success is a collection of problems solved.*"[8] If you can solve critical problems, you will always find work and it can be fascinating and exciting.

When Eve began her career, she thought she wanted to be a research scientist. With time, she felt she was learning more and more about less and less. Then she realized that if she took on technical service work, her work was just as interesting, and the business units would gladly increase funding if she was producing results that solved major problems. Eve realized that what she really enjoyed was making a large positive impact by solving serious problems. She ignored her narrow technical discipline and taught herself whatever was needed to solve the problem at hand and/or partnered with people whose skills complemented hers. In the next phase of her career, when she moved into management, she was eager to take on a wide range of assignments, because she knew she could quickly educate herself and become a problem solver. It helped her also, to have wonderful mentors along the way, and they aided in making her realize her value amid criticism and other difficulties. On top of that, her experience made her a mentor, who asked younger professionals probing questions, to help them envision what kind of future they wanted for themselves, and what skills they already had to achieve their goals.

---

[8] Daum, Kevin, 18 Inspiring Quotes from I. M. Pei, April 26, 2017, https://www.inc.com/kevin-daum/18-inspiring-quotes-from-im-pei.html, as of December 15, 2021.

Maria Angela recognized she was a problem solver when she was given the opportunity to lead teams. Her strength lies in collaborative leadership. Although she suffered severe bullying in elementary school because of low self-esteem due to her thick glasses and orthopedic shoes, she blossomed as an adult, leaving behind her fears and shyness, to enthusiastically inspire and motivate teams. She enabled her teams to be enthusiastic about themselves and their objectives. When women needed empowerment, she created a network. When operators didn't trust new technologies, she launched a pilot project, so that all the stakeholders could ask questions and make modifications. Her insights grew with time, and she appreciates the guidance she received from multiple mentors over the course of her career. They helped her make difficult decisions about changing jobs and moving to another continent to advance her career. Mentors were pivotal for her. Her family support group was central to her analysis of alternatives.

Maria Angela started her shift towards sustainability in 2016, when she became aware of the United Nations (UN) Sustainable Development Goals[9] (SDGs) through a contact with the Kuwait UN International Organization for Migration Office[10] (OMI). When she saw the colorful icons representing each of the seventeen goals hanging in the atrium of the UN building in Kuwait, she decided to learn more about sustainability. She educated herself about sustainability with a focus on a global scope and networking. Her initiative led to the creation of the Geophysical Sustainability Atlas,[11] that maps geophysical techniques and methods to each of the seventeen UN sustainable development goals. This atlas created opportunities for Maria Angela, who also engaged in two certification processes, one with the Institut Francais du Petrole, (IFP) and the other with the University of Cambridge in the UK. This prepared her to shift to an entrepreneurial consulting phase. She is pleased with her new career path that was initiated by securing residency by merit in the USA and moving to Houston in late 2020 when Kuwait eliminated the pandemic travel restrictions imposed on expats.

We crafted the following diagram to illustrate our vision of what works in each phase or cycle of the economic life of industries and hence, employment (Fig. 2.1).

---

[9] https://sdgs.un.org/goals.

[10] https://www.iom.int/countries/kuwait.

[11] https://library.seg.org/doi/10.1190/tle40010010.1.

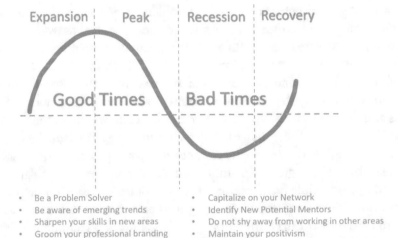

Fig. 2.1 Where you should focus your attention in good and bad times of economic cycles that occur in all sectors

3

Human Nature

While business cycles train us to expect and prepare for change, human nature is resistant to change. Literature from multiple cultures illustrates enduring human characteristics that we should keep in mind while dealing with other people in our lives. We all see the world through our own eyes and most of us place a high priority on protecting ourselves and our loved ones.

3.1   The Origin of the Word Mentor

In the Greek tragedies of the fifth century B.C. by Sophocles, Aeschylus, and Euripides, the principal character, usually a person of importance with outstanding personal qualities, fails disastrously because of a combination of personal flaws and circumstances beyond their control. Those ancient Greek plays still resonate with us, because despite the tremendous changes in technology and culture, inside all of us, primitive emotions still lurk.

The word "mentor" comes from the Odyssey.[1] Mentor was the Greek goddess Athena in disguise. As Mentor, Athena served the teacher of Odysseus' son. Real life mentors are not gods like Athena. They are human beings who attempt to share their experience and their advice, challenging their mentees with options and questions. Like all human beings, they are

---

[1] https://mentoringreece.com/why-mentor-who-was-mentor/.

E. Sprunt and M. A. Capello, *A Guide to Career Resilience*, https://doi.org/10.1007/978-3-031-05588-1_3

not perfect, and their emotions, biases, and personal goals may compromise their recommendations and diminish their value as advisors.

3.2   Elizabethan England

Shakespeare's plays remind us of enduring human emotions that drive behavior just as much now as they did during the reign of the first Queen Elizabeth of England. Matt Ridley's article in the Wall Street Journal[2] on November 5, 2011, summarized why we still enjoy Shakespeare's plays and must manage a wide range of human emotions at work

> *"You can't change human nature." The old cliché draws support from the persistence of human behavior in new circumstances. Shakespeare's plays reveal that no matter how much language, technology and mores have changed in the past 400 years, human nature is largely undisturbed. Macbeth's ambition, Hamlet's indecision, Iago's jealousy, Kate's feistiness and Juliet's love are all instantly understandable."*

Both in our personal and our professional lives, we must learn to get along with many difficult people and recognize that our behavior can contribute to interpersonal friction.

3.3   Latin America

Latin American literature also has famous authors whose heroes' fates are shaped by immutable human behaviors. A dominant theme in the novel, **One Hundred Years of Solitude**, by the Colombian author, Gabriel García Márquez is the inevitable and inescapable repetition of history. This novel, which was published in 1967, has been translated into 46 languages and has sold more than 50 million copies. The 1982 Nobel Prize in Literature was awarded to Gabriel García Márquez *"for his novels and short stories, in which the fantastic and the realistic are combined in a richly composed world of imagination, reflecting a continent's life and conflicts."*[3]

Our challenges in working together also include taking into consideration differences in cultural behavior. What one culture may consider appropriate, may be a serious insult in another. More and more, we interact with people

---

[2] https://www.wsj.com/articles/SB10001424052970204528204577011681658907746,   as   of November 5, 2011.

[3] https://www.nobelprize.org/prizes/literature/1982/summary/.

from a wide range of cultures, and we may inadvertently offend each other. We should remind ourselves to be slow to take offense, because the other person may have no ill will towards us and may just come from a different culture.

3.4   Mentors and Sponsors Are Human

Mentoring is popular with a wide range of organizations, including educational institutions and businesses. Professional groups strive to provide additional professional training and/or coaching for their members. Everyone recognizes that having access to expert advice enhances an individual's ability to succeed. However, like other aspects of life, mentoring depends on trust and involves risks and rewards. Sharing our aspirational goals and our fears enables a mentor to provide more targeted advice, which goes beyond broad platitudes, but also makes us vulnerable if the mentor misuses information we have revealed.

We seek as mentors, people who have achieved success in careers of interest to us. Often, they have overcome many obstacles to reach their goals and we want to learn from them how we can be similarly resilient and successful. If they have time, successful people can generously provide valuable advice.

However, we need to recognize that all people suffer from human failings and that to thrive, we must devise ways to leverage the good in everyone and minimize the adverse impacts of the bad.

Part II

Career Resources

To make a prairie it takes a clover and one bee,
One clover, and a bee,
And revery.
The revery alone will do,
If bees are few.

—Emily Dickinson

4

Mentors

Current wisdom is that everyone needs at least one mentor to achieve success in life and work. Employers often assign one or more mentors to inexperienced, recently hired, or junior employees. Many professional groups and organizations offer a variety of mentoring programs to students and members. Below we share what you need to understand about mentoring and different mentoring relationships to gain as much benefit as possible.

In a pure mentoring relationship, the mentor offers advice in response to queries, but the mentee is under no obligation to accept or act on that advice. For example, you could contact someone who is successful in an occupation of interest to you and arrange a time to chat. Another example is a discussion arranged by a company for their young professionals with more experienced ones, in a structured format, such as monthly encounters. A third example is a professional association that launches a mentoring program between people who don't know each other and do not share an employer but are seeking advice about career progression. Mentoring can be a one-time event, or a series of sessions repeated many times. Let's analyze some types of mentors and their pros and cons.

© The Author(s), under exclusive license to Springer Nature Switzerland AG 2022
E. Sprunt and M. A. Capello, *A Guide to Career Resilience*, https://doi.org/10.1007/978-3-031-05588-1_4

4.1 Types of Mentors

4.1.1 Supervisor Mentors

When a mentor is also your boss or supervisor, the mentor's guidance is not "pure" in the sense that if you as the mentee don't follow the advice, there may be adverse consequences for both you and your supervisor. New employees may be assigned mentors to help them with different aspects of their life at work including cultural (the expected behaviors in the organization) and technical (job specific). This type of relationship is in effect a trainer-trainee situation, with the subordinate's behavior potentially impacting the career of the more senior person.

Adhering to organizational norms is important to achieving success with many employers. Guidance on how a junior person should behave in the organization helps that person to achieve success with that employer but may or may not be the best way to advance elsewhere. Organizations have their quirks, and it is important to recognize which behaviors are widespread and which are part of a particular group's culture.

Some organizations value and encourage challenging the status quo and suggestions of how to improve processes. Others are not receptive to junior personnel with limited experience questioning long time practices or their leaders. Organizations may judge both the trainer and the trainee on how well the junior person mimics and adheres to the organization's culture.

4.1.2 Technical Mentors

Technical mentoring involves tutoring the junior or less experienced person on how to do their job. The employer judges the technical mentor on how well the junior person performs and judges the junior person on their acquisition of the skills required for their assigned tasks. Some professions, including music and medicine, require dedicated one-on-one technical mentoring. In most professions, these mentors supplement technical training.

4.1.3 Outside Mentors

Many times, people have mentors in other organizations or areas of expertise. These mentors may be found through professional societies, friendship or family connections. These mentors are especially valuable, because they do not have a vested interest of their own in your progress inside your organization. They can offer you their insights and guidance from what is generally an impartial and fresh point of view, independent of the politics that your own work organization imposes.

4.1.4 Professional Mentors

Currently, a variety of companies offer mentoring programs that are contractual agreements with professional consulting firms or other organizations that provide mentoring as a service, with internal or external experienced individuals who fulfill the role of mentors. In addition, given the relevance of mentoring in the current business environment, to accelerate the careers of young professionals, not-for-profit organizations including governmental associations and professional societies also offer mentoring programs with mentors and mentees residing in many locations, cultures and time zones. Although the global scale can pose challenges for this type of online mentoring, these relationships can provide important insights and guidance not otherwise available (Fig. 4.1).

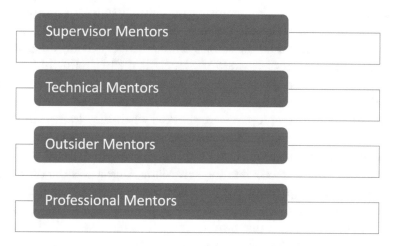

Fig. 4.1 Types of mentors you may encounter in your career

4.2   Academic vs. Corporate World: Eve's Experience

When Eve began work, she didn't realize the company culture differed from her universities' rigorous academic culture. Seminars in the universities included sharp questioning of the speaker. Eve naively thought that was the way it was everywhere and quickly gained a reputation for being overly aggressive. Colleagues told her the company culture was "praise in public, punish in private." In technical seminars, only encouraging questions asking for more details were acceptable. If someone suspected a serious technical flaw, the acceptable way to address it was in a discrete, private conversation. Eve subsequently observed that large and expensive projects with serious technical flaws were quietly and diplomatically shelved, with especially large and expensive failed projects swept under the rug in a reorganization.

4.3   Cultural Challenges: Maria Angela's Experience

Maria Angela's bosses made mentoring on cultural styles a priority because they found managing her independent and creative style challenging, especially in meetings or public gatherings. She remembers vividly an experience in the operational offices in eastern Venezuela at the beginning of her career. The giant El Furrial Field was in the early stage of development, with priority on accelerating drilling. The boss, who was a micromanager, called a meeting of all personnel. The employees were shocked when they learned the reason for the meeting—too much paper was being used for photocopies. The boss had the only photocopier which was available to everyone placed in a locked room to which the secretary had the key. Maria Angela was the only person who dared to speak up. She asked if it was possible to distribute the paper to the five coordinators to determine which group was using the most paper and to evaluate how to reduce that group's usage. The big boss replied it was not possible, because "*they had already made a copy of the key to the room.*" Multiple supervisors advised Maria Angela that it was not good or wise to challenge authority in a general meeting. Maria Angela soon learned the ropes, but still thought it was better to speak up than to accept lousy decisions.

Later, as a Halliburton employee, Maria Angela realized people were very defensive about their work. People did not have the informal relationships and conversations she enjoyed in the Venezuelan National Oil Company, PDVSA. She had to adapt to a corporate culture in which people were afraid of losing their jobs. Misinterpretation of anything you said could lead to unintended repercussions. Therefore, in all her communications, whether in person or by email, Maria Angela carefully avoided statements anyone might take as a personal or professional criticism of their work. Her goal was to encourage collaboration and integration between teams. In retrospect, she realized in the National Oil Company (NOC), employment was more secure, and people felt freer to express themselves. In contrast, in the international service company, people worried about their continued employment, and performance was of paramount importance. It is essential to understand what is motivating behavior in an organization. When employment is secure, people feel freer to raise their concerns than when layoffs and terminations are frequent.

In the Middle East, Maria Angela was a maverick. She was the first female technical consultant from Latin America in her company. The executives were familiar with the style of American and British people, and most of the employees were from India. Maria Angela was distinctly different. As is common with Italian Venezuelans, she is talkative, outspoken and direct. Her bosses and colleagues explained to her in private meetings that the distance to power was important in their corporate culture. Maria Angela remembers one of her first meetings with a Deputy CEO Exploration about training. In the meeting, after all groups shared their yearly plans, he asked for feedback, ideas and comments. Maria Angela raised a few comments, and she noticed her immediate boss signaling to her across the table to look at her mobile phone. He had messaged her that before speaking she needed to say, "*May I speak to you, as I have a comment.*" This shocked her, but she needed to be aware of it, because it was part of the corporate culture to request permission before talking.

In her volunteer work with the Society of Exploration Geophysicists (SEG), Maria Angela noticed a distinct organizational culture. Academics on the Board of Directors warned her, "When *you speak, it seems you control the truth based on your experience and talent. But most of us come from academia, and we first acknowledge the participation and ideas of those who have already spoken before we express our own perspective.*" That board member mentored Maria Angela. With his guidance, she better understood the culture of SEG and learned how to modify her behavior to avoid miscommunication.

4.4   Challenging Mentors

When a junior person questions the value of their assignment, it can be very awkward for their mentors. Eve made that mistake when she left the academic environment and joined Mobil's research laboratory. As a new employee, she was responsible for the laboratory in which technicians measured the speed of acoustic waves through core samples from oil fields. The project's goal was to understand the seismic velocities in a specific area with different fluid mixtures. When Eve learned, the technician had smoothed the ends of the samples, fixed cracks, and plugged holes with multiple kinds of epoxy, she was worried the results would provide no real insights into the seismic velocity of the rocks in the subsurface. Her boss's reaction was that the samples were the best they could get. Despite flaws in the experimental process, Eve capitulated. After providing the requested tests, Eve figured out how to avoid being stuck producing what she believed were potentially misleading results but realized that challenging mentors is not always the best alternative, because they do not welcome it.

People we interviewed also shared problems arising from having their boss as a mentor. One woman described how even though her boss was much older than she was, her boss *"felt threatened to such a point that everything I was suggesting, she took as an opportunity to go against me."* Fortunately, her supervisor's boss supported her and explained that strong contributors like herself often are attacked by people who see them as a threat to their own success. Her conversations with her boss's boss helped her gain a new perspective and develop ways to cope with the situation.

Other women also reported having trouble with female supervisors, who were supposed to mentor them. One described her experience. *"She was very nice but started turning on me and putting me down. She went from sweet to nasty within a month. She would dismiss any project I suggested, but if a male suggested it, it was fine. Anything I said was wrong. She would not allow me to do anything. It was very, very difficult."*

Managers may not be the best technical experts. But a fresh graduate might think they are, as Maria Angela experienced in the Middle East where they revere the bosses in a hierarchical and high power distance culture. It is difficult for recent graduates to realize that their boss may or may not be able to function as a technical mentor. They assumed everything their boss said was technically correct, even if it concerned geosciences, and the boss was a mechanical engineer with no experience at all in geosciences.

4.5   Impact of Culture on Mentoring

Maria Angela had many excellent technical mentors throughout her career, beginning at her university, where professors ignited in her a passion within her physics studies for geology and geophysics. She opted to take a minor in geophysics. At Lagoven, experts in sequence stratigraphy from the French company, TOTAL, mentored Maria Angela. The mentoring went both ways. They knew the seismic techniques, and Maria Angela knew the geology and geophysics of the field. This experience made Maria Angela value aspects of technical mentoring that vary with national culture: expectations from the mentee, commitment of the mentor to the learning process of the mentee, and a personal connection between the mentor and mentee.

In Venezuela, the expectations about what the mentee needed to achieve were less strict than with a French mentor. The French mentors were very formal and obviously held accountable by their bosses. In contrast, Venezuelan technical mentors offered their support, rather than a formal mentoring program. Also, technical mentors in Venezuela sought more of a personal connection, even at the family level than the French ones.

Maria Angela has served as a technical mentor and organized technical mentoring programs for her employees and clients while in Halliburton and in KOC (Kuwait Oil Company, the National Oil Company of Kuwait). In recent years, Maria Angela has developed an influential presence online that has attracted requests for online mentoring from people in multiple countries.

We must take the preferences and priorities of both mentors and mentees into consideration, even for purely technical mentoring. Agreement on the objectives, timeframe, and organization of mentoring sessions in technical and professional settings is not only desirable, but essential to set expectations and create a healthy relationship.

4.6   Academic Advisors

Many academic sub disciplines and specialties are relatively small groups, so all the key players know each other. Think about who will be your advisor when selecting a graduate school. Your thesis advisor for doctoral studies is essentially your boss.

Often the doctoral student depends on their advisor for both personal income (stipend/fellowship) and research funding. Even if the student has an independent fellowship, their advisor is usually still the source of the research funding. Ultimately, the advisor's approval of your thesis is essential. This relationship can be fraught with tension. The student and the thesis advisor may have different opinions about whether the thesis research is sufficiently complete, and the thesis advisor is a critical sponsor for future employment.

The student's goal is to complete their thesis in a reasonable time and to get acceptable employment at the end. A lot can go wrong. Some of the most exciting work can be associated with new, untenured faculty. However, if the faculty member, for whatever reason, fails to get tenure, that may completely disrupt their students' work. When considering where to go for graduate studies, research the reputations of the faculty members you are considering for your thesis advisor.

When Eve was finishing at MIT, she was lucky to get a National Science Foundation Fellowship, so she had funding for tuition and a stipend for three years. MIT, Stanford, and CalTech accepted her for their doctoral programs. The professors with whom she would have worked at MIT and CalTech had tenure. Eve's MIT master's thesis advisor, Bill Brace, said the professor she was targeting as an advisor at Stanford, Amos Nur, would get tenure. Prof. Brace made Eve's decision easier when he said if she didn't like Stanford, she could be back at MIT in six months. Prof. Brace went even further and introduced Eve to Prof. Nur.

Eve's experience illustrates how thinking ahead about potential mentors is important for success in academia. As a student, do not remain silent about potential alternatives and outcomes of your research programs because you have the right to ask. Actively gathering as much advice and information as possible will greatly facilitate your way ahead.

Another major consideration for graduate students is the selection of their thesis committee. Some faculty members may be attempting to prove a theory of their own or need results to support their research programs. If your experimental work does not support that person's theory or research program, they may request more and more experiments. If you run into that situation, your best alternative may be to have that faculty member replaced. Trying to disprove a professor's pet theory can be a long and frustrating experience.

If you have your thesis advisor's support, replacing one member on your thesis committee is usually not a major problem. Eve had one member of her committee replaced when her experimental results repeatedly failed to support his theory, and he kept requesting more experiments in a desperate attempt to get support for his theory.

Undergraduate thesis advisors may not always make the student's best interests a top priority. At the end of her freshman year, Eve selected geophysics as her major and chose a seismology professor as her advisor. She wanted to do research, and he assigned her to work with a Turkish research associate who spoke very little English. All that academic year, she collected data and when she asked about going to geology summer field camp, her advisor recommended she spend the summer working for him and continuing to gather data.

When the advisor and the research associate published a paper based on the data Eve gathered and analyzed, they didn't even acknowledge her. In contrast, a male contemporary of Eve's worked directly with the same professor and co-authored a paper with him. Fortunately, Prof. Brace, a rock mechanics expert, invited Eve to work with him the following year and she co-authored three papers with him. Unfortunately, Eve did not speak up quickly, and silence was not the answer to getting the training and recognition she needed from the seismology professor. Because of the seismology professor's bias, Eve became an expert in rock mechanics instead of seismology.

In hindsight, all turned out well, because a few years into her career, Eve slid from rock mechanics into petroleum engineering and never worried about what her official technical discipline was. Her real passion was for solving technical problems and using whatever technology was appropriate to find an answer. She learned to ignore what many other people perceived to be barriers between technical disciplines.

Eve's change of advisor and university was a major decision. Besides not being silent, many times we must shift gears completely to succeed. One huge benefit of earning her degrees from more than one top university was that she became a member of two powerful networks. Another benefit was that she learned that although both universities, MIT and Stanford, are very highly regarded, they considered different things to be important and had different cultures.

Maria Angela had several academic advisors during her Master studies at the Colorado School of Mines, including Michael L. Batzle, Thomas L. Davis, and Ramona Graves. They saw in Maria Angela a committed student and guided her even after graduation. Prof. Batzle sponsored Maria Angela for many other opportunities, opening doors with his connections in the oil industry, which proved instrumental in advancing several of her projects at INTEVEP.[1] Professor Graves continues to be a treasured mentor and sponsor of Maria Angela. Academic advisors can be lifelong sponsors, who "open doors for you." If you prove your talent and value to them, they may continue to provide valuable guidance throughout your career.

---

[1] http://goldmercuryaward.org/laureates/instituto-tecnologico-venezeolano-del-petroleo-intevep/ as of December 15, 2021.

Maria Angela had academic advisors in her undergraduate studies in Venezuela and in her graduate studies in the United States. In her undergraduate courses, almost all her professors were dedicated, exceptionally knowledgeable and committed to making Maria Angela a researcher in physics. However, she chose to get a minor in geophysics. The geophysics professors differed from those in the physics department in that some of them spent only a small fraction of their time on teaching.

Studies with a professor of seismic interpretation provided a memorable lesson for Maria Angela. He was an associate professor who was also working in industry, so he could dedicate very little time to providing explanations to the students and assigned exercises about seismic interpretation with no guidance. Maria Angela could not bear the situation and requested a meeting with the department head to make a formal complaint, knowing that she was taking an enormous risk, especially regarding her grade in that subject. To her surprise, the department head listened carefully to all the issues Maria Angela raised, asking for specific examples. After the department head discussed the situation privately with the professor, Maria Angela noticed a significant improvement. She was fearful of reprisals but received a merited high grade. Afterwards, she realized that all the students in the class had been struggling and were unhappy, but no one else dared to speak up. Indeed, it takes courage.

Challenging bad mentoring is recognition that silence cannot solve anything. After a difficult conversation, things will never remain the same. They will get better or worse. If you are clear and factual about the problems and your objectives in this type of conversation, the outcome is usually positive.

Life is never without risks. Staying silent has different risks than speaking up. You are more likely to succeed in life if you phrase your requests for change in a positive way and show how the changes you want will benefit not only yourself, but other stakeholders.

4.7   Selecting a Mentor

Mentoring has become so ingrained in career development, most people now have one or more designated mentors. However, being assigned a mentor does not automatically make the relationship successful or beneficial. When mentoring works, it is extremely valuable for the mentee, but in many cases, there isn't a good match between mentor and mentee.

As in other types of relationships, many factors can contribute to success or failure. It may be necessary to try several mentors, before you find one who is good for you. When you find a great mentor, it can be life changing for the better. The secret is not whether the mentoring program follows formal guidelines or is conducted face-to-face as opposed to remotely. Beneficial mentoring may continue over an extended period or occur during a fleeting and perhaps long-distance sessions. The secrets to success include establishing rapport and the mentor possessing knowledge and or experience that is of value to the mentee.

Never underestimate the tremendous value that having a variety of mentors can bring to your career decisions. The mentor is not the decision maker, you are, but they can provide extremely valuable insights for you to take into consideration in making your decision. Multiple mentors can provide a range of different and insightful perspectives.

Individuals and organizations often confuse mentoring and teaching. A mentor is a guide who helps you figure out which path you want your career to follow and what skills and knowledge you will need to be successful. A good mentor asks you probing questions and provides introductions and suggestions on how to identify and explore alternatives. They may also offer recommendations about other people you should contact to further explore your preferences and aspirations.

A valuable mentor will ask you difficult or sometimes uncomfortable questions that push you to more deeply and critically examine your preferences for the future. You should prepare for your mentoring sessions and be open to inquisitive interactions. Although some mentors may provide technical tutoring, that is not the classic role of a mentor. Think of a mentor as someone with whom you discuss different potential career paths and what you will need to be successful in your career. Mentors will not necessarily teach you the technical skills you will need.

For mentoring to be successful, the mentee must reveal their hopes, dreams, and concerns. Most people feel vulnerable when they reveal their weaknesses as part of a discussion about the gaps they seek to fill. Knowledge of someone's weaknesses can be used against them.

Trust is essential in a successful mentoring relationship. Merriam-Webster defines trust as "belief that someone or something is reliable, good, honest, effective, etc."[2] We may enter a new mentoring relationship with some trust of the mentor based on their reputation, but deep and lasting trust develops

---

[2] https://www.merriam-webster.com/dictionary/trust as of December 15, 2021.

from our personal experiences with the mentor and the advice and coaching that they provide. That type of genuine trust takes time to develop.

For a valuable mentoring session to occur, both people must trust each other. If you become suspicious that your mentor is not acting in your best interests, stop sharing your secrets and find another mentor with whom you feel freer to reveal your hopes and dreams. Do not continue to divulge sensitive information if you suspect that it may be used against you in any way.

Society works as well as it does, because we are not all identical clones, and we learn to coexist, valuing everyone's contributions. People have different styles, priorities, strengths, preferences, values and risk tolerance. The perfect job for one person is not necessarily the best for another. Strong mentoring relationships grow from shared values and preferences. The best advice for one person can be woefully wrong for another. When a person has multiple mentors, they can evaluate a wider range of opportunities. Ultimately, you are responsible for making the decisions for yourself.

4.8   Mutual Mentoring

Some of our best mentors can be our colleagues. They may understand better than we do the organizational culture and personalities of key people. Often, we can be more candid about our hopes and fears with trusted colleagues. When we need guidance about dealing with a troublesome person within our organization, these colleagues can be our best source of advice. They can make a better assessment of the situation than a spouse or confident outside of the organization, and provide actionable, practical advice.

Sometimes, the best mentors are not the most obvious. Eve had a colleague who was often more of a hindrance than a help tell her, "You'd better hope you never have a Middle Eastern man as your boss." That narrow-minded American man didn't realize that Eve's best friend and confidant within the company was a devout Muslim man from Syria. Eve's confidant and his wife have been her very close friends for decades now. Eve often received much better treatment from Middle Eastern men than she did from American men. Eve's pivotal sponsor during her early career was a Palestinian Christian. She learned early in her career that some men from the Middle East (both Muslims and Christians) were more supportive of her than many men raised in the United States. You shouldn't prejudge someone by their religion or ethnicity or place of origin.

Maria Angela received extraordinary support from a male colleague at KOC when he explained to her the invisible lines of power that were not obvious from the organizational chart and official position titles. He also explained why some individuals in different groups or departments were instrumental for decision-making processes in the company. In addition, he took the time to explain the different religious and cultural nuances of Kuwait that impacted the work environment. He enabled Maria Angela to become a politically savvy influencer who could leverage her knowledge to advance initiatives on the corporate scale. From experience, she realized that colleague or peer-mentoring is one of the most important ways to understand company culture.

Maria Angela has enjoyed mentoring relationships with a variety of men and women with whom she feels at ease sharing her aspirations, thoughts and/or challenges. The communication with some is face-to-face and with others continues to be remote. She has learned to distinguish collaborative, helpful peers from those peers who treat their colleagues as competitors. If peers care about you, they engage in mentoring. Competitors do not. You can detect the difference when you engage in a conversation. Supportive peers ask inquisitive questions to help you better understand an issue. Also, a well-intentioned peer may offer options, contacts, and introductions to help you achieve your objectives. Competitors avoid sharing valuable information. Beware of bad or misleading mentoring from someone who seldom divulges valuable information in a timely manner.

An important mentor for Eve was a female colleague, who was about the same age as Eve and at the same managerial level, but who had been with the company much longer. When Eve complained about being bullied by a high-level female executive, her colleague revealed that that female manager had bullied almost every woman of ability, including herself. That knowledge made it much easier for Eve to figure out how to handle her problem. When we share our problems with our friends, it is amazing how much useful information we learn.

4.9 Professional Society Mentoring

Many organizations provide mentoring services. These services take multiple forms that in the absence of pandemic restrictions about face-to-face contact include speed mentoring at conferences. New programs (2021) include

Female Leaders in Energy[3] (FLIE), which receives funding from the United States Bureau of Energy Resources[4] (EMR) and aims to advance the professional development of early- to mid-career level women working in energy sectors across southeast Asia.[5] Others, like Ally[6] (a private, for-profit initiative) and Lean-In (not for profit) pair people for mentoring in structured approaches for extended periods of time.

The Society of Women Engineers (SWE)[7] pairs members for one-off video conference sessions. Eve found the mentoring through SWE to be more rewarding from her perspective as a mentor because the mentees select who they want as a mentor from descriptions provided by the mentors. She found pairings by other organizations which were based on information submitted by mentors and mentees but did not involve either the mentee or the mentor in the matchmaking to be less successful.

Maria Angela thinks Lean-In Kuwait[8] has an excellent framework for mentoring. In other countries including the USA, she found the Lean-In mentoring programs did not work as well. She enjoys the FLIE mentoring program, because she considers the materials provided truly guide both the mentors and mentees towards a successful liaison.

The SEG (Society of Exploration Geophysicists) also provides mentoring support through the society's Women's Network.[9] Eve was the founding chair and Maria Angela was her successor as chair.

@GeoLatinas,[10] an organization which supports online interactions and connects Latinas all over the world, currently has nine thousand members and continues to focus on peer-mentoring of women, primarily students and recent graduates.

Many organizations in academia also provide mentoring programs. For example, in 2021, the Colorado School of Mines launched a mentoring program[11] using alumni volunteers as mentors. Based on her service as a mentor in this program, Maria Angela ranked it as one of the best in which she has been engaged because of the resource materials provided on the issues

---

3 https://usea.org/program-categories/flie as of December 15, 2021.

4 https://www.state.gov/about-us-bureau-of-energy-resources/ as of December 15, 2021.

5 https://usea.org/program-categories/flie, as of December 11, 2021.

6 https://allyenergy.com/about/#:~:text=We%20give%20professionals%20a%20community,overlo oked%20career%20on%20Mother%20Earth as of December 11, 2021.

7 Society of Women Engineers (swe.org), as of December 11, 2021.

8 https://leanin.org/circles-network/kuwait-kuwait-city#! as of December 15, 2021.

9 https://seg.org/News-Resources/Womens-Network as of December 15, 2021.

10 https://geolatinas.weebly.com/ as of December 11, 2021.

11 https://www.mines.edu/mentoring/ as of December 11, 2021.

most often raised by the mentees. The students in this mentoring program usually are seeking advice about future employment opportunities, salary negotiations, and long-term career goals.

4.10 Casual Mentoring

Mentoring doesn't need to be formal. Casual voice or video conversations by telephone, email, video conference or over a cup of coffee or lunch can provide valuable career guidance. What matters is that all the participants feel comfortable discussing sensitive issues. Trust that the other person will not misuse any sensitive information is an essential element.

If complete confidentiality is important, make sure you make that clear at the beginning of the conversation. If the mentor hesitates, be sure to clarify expectations before continuing your discussion. If the mentor declines to guarantee complete confidentiality or you begin to suspect that confidentiality will not be maintained, switch to a non-confidential topic and find another person who is willing to speak to you about the issue and agrees to keep your discussion confidential.

5

Sponsors

People often mention mentors and sponsors in the same breath, but there are critical differences. Mentors are sources of advice and could be any person with the right experience. In contrast, a sponsor is a person with the power or influence to intervene on your behalf to improve your chances of being selected for a highly competitive and desirable position. That gives sponsors much more power, which means when the relationship goes sour, they can be much more destructive.

5.1 Role of a Sponsor

Your sponsor is a person in a position of power and/or influence, who supports your candidacy for a position by providing a recommendation or endorsement and using their influence to persuade others to support your selection.

Acting as a sponsor is risky. If the future performance of the sponsor's protégé doesn't meet expectations, that failure reflects badly on them. By endorsing a protégé, sponsors are investing a portion of their reputation in them. The value of most sponsors' recommendations depends on the track record of the people they have supported. If a sponsor recommends someone who performs poorly, their endorsements lose value.

If many qualified people are competing for a position, a powerful sponsor's recommendation may be essential, and there is potential for abuse and/or

© The Author(s), under exclusive license to Springer Nature Switzerland AG 2022
E. Sprunt and M. A. Capello, *A Guide to Career Resilience*,
https://doi.org/10.1007/978-3-031-05588-1_5

corruption. As the old saying goes, "It's not what you know, but whom you know". Lord Acton, one of the most illustrious historians of nineteenth-century England, wrote in a letter to Archbishop Mandell Creighton dated April 5, 1887, *"Power tends to corrupt and absolute power corrupts absolutely. Great men are almost always bad men, even when they exercise influence and not authority: still more when you superadd the tendency or the certainty of corruption by authority."*[1]

Fortunately, few people have absolute power over anything. There are many good people who serve as sponsors, but there are also those who groom and then abuse their victims, gradually enticing them to be susceptible to their misdeeds.

5.2   Mentors Who Are also Sponsors

One example of a mentor who often also serves as a sponsor is the faculty member who supervises a Ph.D. thesis. Endorsement by your thesis advisor is essential if you wish to get an academic position after completing your studies. The stronger your advisor's network, the greater your chances are of securing a coveted position.

5.3   Finding a Sponsor

People don't just "wear a single hat." They may fulfill multiple roles in your career. Someone from whom you seek advice may subsequently become an important sponsor. While you can ask almost anyone for advice, a stranger is unlikely to endorse you for a highly competitive position. However, asking someone for advice can be the entry into a relationship in which they subsequently serve as a sponsor.

Another way to find a sponsor is to say "Thank you" to someone who has helped you. Early in your career, someone may, without your knowledge, have recommended you for a position. If you find out someone has helped you, say *"Thank you."* It may change your life!

---

[1] https://www.englishclub.com/ref/esl/Quotes/Politics/Power_tends_to_corrupt_and_absolute_power_corrupts_absolutely._2652.php#:~:text=%22Power%20tends%20to%20corrupt%20and%20absolute%20power%20corrupts,corruption%20by%20authority.%22%20Lord%20Acton%20%281834-1902%29%20English%20historian viewed on February 2, 2022.

5.3.1 Eve's Experience Finding a Sponsor

When Eve began work in the petroleum industry in the 1970s, there were very few women. Many men thought it was no place for a lady, but a few did. Eve, whose degrees were in geophysics, found herself invited to serve on the Geology and Geophysics subcommittee for the Society of Petroleum Engineer's (SPE) annual meeting. She missed her first meeting because it conflicted with her vacation. The next year, the oil price plunged, so the chairman of the committee took a job in Saudi Arabia and asked Eve to represent the committee at a meeting of the overall program committee.

When she went to the program committee meeting, Eve had to select the papers for a session at the annual meeting, because the Geology and Geophysics subcommittee had not met to determine a program. The chair of the committee had not even shared with the committee members the abstracts that had been submitted. Eve had to work fast. She read the submissions, selected the ones to be presented, and agreed to chair the session at the annual meeting. Although she thought that was the end of her responsibilities, SPE asked her to chair the subcommittee the following year. After a friend at Mobil informed Eve that the company's senior scientist, Dr. Aziz Odeh, had volunteered her for the subcommittee, she went to thank him. Later, she learned Aziz had volunteered many people to serve on SPE committees, and she was the only one who had ever thanked him.

At a time, Mobil's management thought it was inappropriate to have a younger woman supervising older men. Eve never advanced into a management position at Mobil, but without Eve's prior knowledge, Aziz repeatedly endorsed her for roles of increasing responsibility within SPE, and she became the first woman to serve on SPE's Board of Directors.

After each move up the SPE hierarchy, Eve visited Aziz, often to find out how to get Mobil's permission to accept the new volunteer position. He advised her, *"tell them you will do as much of it on your own time as you can and do that. It will get easier with time."* That was tough advice for the mother of two young children, but Eve followed his recommendation, and he was right. With time, Eve learned she could leverage her SPE leadership positions to enhance her status with her employer.

Eve's involvement with SPE played a huge role in shaping her career and her life, and it all started with her saying, *"Thank you"* to someone she barely knew. Giving thanks is important, especially if you realize a person has acted as a sponsor for you, opening doors or supporting you for roles or opportunities. Thanking a person creates a special bond, based on trust, but also in humbleness, because you are not the only ones deserving opportunities, and it is a privilege to be selected for desirable or influential roles.

5.3.2   Maria Angela's Experience

Maria Angela did not know she had powerful sponsors during her tenure in PDVSA. In hindsight, she realized she was the first woman in Venezuela (and probably first Latin-American) to be selected to be a field supervisor of seismic crews. That came not only because of the regular training of her company, Lagoven[2] (former Exxon, Creole Petroleum Corporation Venezuela), which demanded field experience for all geoscientists and engineers in Exploration departments, but also because of early identification of her talent by her first manager. He sped up the timing of her transfer to field operations, so she began supervising a field crew just seven months after starting work.

Maria Angela believes the manager's sponsorship resulted from her state-of-art use of computers in technical presentations to support his recommendations. By working with the manager on these presentations, he got to know her better. Another way to gain a valuable sponsor is to provide technical support to executives.

Later in her career, Maria Angela discovered the importance of asking for what she wanted. When pregnant with her first child, she believed she could not risk the long commute to her home in a nearby city. She asked to talk about "a personal matter" with the General Manager, three levels above her. All the intermediate supervisors worried they had offended her and wanted to speak with her, but she insisted that her issue was confidential, and she had to speak with the big boss. When she spoke to the General Manager, he recognized her courage and not only approved her transfer to INTEVEP, the research center near to her new house, but became a sponsor for her until his retirement.

When you speak up and are honest about your request, you show your courage and knowledge of how to work the system. In Maria Angela's case, she not only achieved what she wanted but also gained an important sponsor she had not previously known.

5.4   Significance of Shared Background

In INTEVEP, Maria Angela had the sponsorship of a director who shared her Italian heritage. In Venezuela, descendants of Italian immigrants are the second largest non-indigenous group. Although Italian-Venezuelans were common in other industrial sectors, they were under-represented in the oil

---

[2] https://www.bing.com/search?q=Lagoven&form=ANNTH1&refig=42e29c6d88b94cf596e6f6a3ec7 48b32, as of December 15, 2021.

and gas industry. When Maria Angela received the highest evaluation in her department, the Director came to congratulate her in her office, without an appointment or advance notice. She believes he was instrumental in her transfer to PDVSA to work in what was to be the first joint venture between the oil company and a university, PetroUCV, and that he wanted to support her, because of their shared heritage.

Potential sponsors may wish to help someone for multiple reasons. Besides shared ethnic heritage, executives may favor those who went to the same university that they did or share other experiences or interests. Executives with working daughters may be more likely to sponsor their female employees than executives with only sons or daughters who are homemakers. Eve's sponsor, Aziz Odeh, had two daughters who were professionals. One was a doctor, and the other was a chemical engineer, with whom Eve later enjoyed collaborating with at Mobil.

5.5   Make Your Sponsors Look Good

Maria Angela spent 15 years working in Kuwait for Kuwait Oil Company. The first sponsor she found was an astute manager who rapidly recognized how Maria Angela could help her and her department. Making your boss look good is the classic way of getting a sponsor.

In Kuwait, Maria Angela's boss connected her with a deputy CEO of the Company. Maria Angela guided, mentored, and coached those two leaders, and helped them engage with professional societies, international organizations, publications, and other companies to showcase their successes. Maria Angela made herself so valuable to these two leaders, they protected and endorsed her even though she was a female foreigner.

After those two leaders retired because of their age, Maria Angela had other sponsors at the C-suite level who continued to leverage her experience and gravitas to support, enhance and progress the company's reputation. More than five years after she lost her main sponsors, Maria Angela learned about angry social media posts in Arabic, which identified her. The posts asked, "*Why don't we have national people able to write articles like this one by Maria A. Capello, or have her prominence?*" and "*Who is this Maria A. Capello, that does so much praising Kuwait? Why do we not have a national person doing this?*"

A Deputy CEO told Maria Angela several people, including himself, protected her and supported her work against these attacks. Maria Angela always praised Kuwait and the Kuwaiti nationals. That praise was not premeditated, but came from her heart, and it paid off.

The lessons learned from this sponsorship experience are that you should support everyone who asks for your help, because you may not be fully aware of the behind-the-scenes politics. Work relationships are multi-faceted. When you deal with powerful leaders, corporate politics and national interests may lead to unexpected complications. Leveraging her extensive network, Maria Angela replaced her retired sponsors with new sponsors outside her department. Networking with a vision of the future is essential for earning the support of sponsors.

5.6   Power of Volunteer Activities

Both Maria Angela and Eve recognize that their volunteer work for professional societies enabled them to find important sponsors and opportunities. No matter where you are in the world or in your career, you can stay connected through volunteer work. When you help others, you help yourself. Volunteer activities showcase your knowledge and abilities and enable you to build a powerful network of contacts and sponsors that extends far beyond your own organization.

6

Role Models

Role models can be very valuable. They don't even need to know you to make a difference. When you see someone like yourself successfully forge a new type of career or make a major career shift, they can encourage you to emulate them.

We are surrounded by potential role models. Look around for people who are enjoying careers or lifestyles of interest to you. You may even try to contact them for mentoring in the form of a few questions about their life and choices. If people are proud of themselves, they usually like to share their wisdom.

Don't make the mistake of thinking role models must be like you. When Eve and Maria Angela began their careers, there were no women for them to emulate. If you find yourself in that situation, look for someone who is doing work of interest to you or getting opportunities that you would like to have.

We often emulate other people's behavior without realizing that we are looking at them as role models. This was true for Eve. A few years into her career, Eve noticed that her male colleagues who had the opportunity to travel internationally were working on technical service as opposed to research projects. Watching them, she realized that she could enjoy higher visibility, greater recognition and travel if she sought technical service projects, and she did.

The first time Eve consciously perceived herself as imitating a role model was about twenty years into her career. She observed a woman, who had been a year ahead of her at MIT, switch from one giant company to another and

© The Author(s), under exclusive license to Springer Nature
Switzerland AG 2022
E. Sprunt and M. A. Capello, *A Guide to Career Resilience*,
https://doi.org/10.1007/978-3-031-05588-1_6

move into a much higher and more satisfying position. Eve had maintained contact with the woman since MIT, and although she wasn't a close friend, she had benchmarked her career against hers. With that woman as a role model, Eve was emboldened to seize an opportunity to switch companies. Taking that mid-career risk worked out very well for Eve.

6.1   Identifying Potential Role Models

Looking at potential role models is like virtually trying different work and lifestyles. Since we often do not know the person well, we don't really know how they feel about their lives. Sometimes people can end up mimicking someone who is a fraud. An example of someone who was admired and copied is Elizabeth Holmes, who in 2003, at the age of 19 dropped out of Stanford University and founded a company that claimed to be able to make a wide range of tests using just a drop of blood from a finger prick.

Holmes became world famous when *Fortune Magazine* put her on their cover and included her in their "40 Under 40 in 2014." *Fortune* explained their choice by writing, "Holmes, whom we put on our cover in June, dropped out of Stanford when she was 19 to found Theranos, a revolutionary blood analytics company. Ten years later, her private, Palo Alto-based company employs more than 500 and is valued by investors at about $9 billion. (She retains more than half the stock, so you can do the math.) Theranos performs blood tests for a fraction of the price charged by competitors and without the use of a syringe."[1]

Eve found Holmes' success hard to explain because Holmes didn't have any known technical expertise. However, at the peak of her success, many women mimicked Holmes' style which in turn mimicked Steve Jobs, the founder of Apple, whose trademark outfit was a black turtleneck shirt. Holmes' reign as a technical wonder woman did not last long after *Fortune Magazine* made her world famous. *The Wall Street Journal* published John Carryou's carefully investigated story about the fraud behind Theranos[2] in October 2015 causing Holmes' career to crash. After years of investigation, legal delays, and a lengthy trial, Holmes was found guilty on four counts of defrauding

---

[1] https://fortune.com/40-under-40/2014/elizabeth-holmes/, viewed January 30, 2022.

[2] https://archive.ph/20160907025454/http://www.wsj.com/articles/theranos-has-struggled-with-blood-tests-1444881901, viewed January 30, 2022.

investors on January 3, 2022. She has not yet been sentenced but is expected to spend time in prison.[3]

If you are unfamiliar with the background of your potential role model and you plan to emulate their behavior, do some research. When possible, arrange to speak with the person and ask them a few questions about their career and what they would do differently in hindsight. Contacting her or him might convert a role model into a mentor or sponsor, who is even more valuable.

6.2 Don't Require Perfection

Someone doesn't need to be perfect to be a role model. We can admire and want to emulate some aspect of their behavior without approving of all their behavior. Since none of us are perfect, we should use role models as examples of what types of benefits certain behaviors make available and what problems they create. Pick the aspect of behavior you are considering emulating and learn from the role model how that impacted their personal life and their career.

We are constantly considering tradeoffs. Observing the impact of those tradeoffs in other people's careers can help us decide whether the rewards are justified by what we potentially lose in making a change.

6.3 Becoming a Role Model

If you are a pioneer, who is working in a new or unconventional way, you may quickly become a role model for others. Eve discovered this shortly after she finished her doctorate and became the first woman to earn a Ph.D. in Geophysics from Stanford. Her husband had not yet completed his degrees, so Eve continued at Stanford as a Research Associate at Stanford. Nine months after her thesis defense, she gave birth to her first child.

When he was thirteen days old, Eve put him in a front baby pack and brought him into work with her, where she shared an office with two men. She strung up a rope to create a modesty curtain for when she breastfed Alexander. Eve brought him to work with her every day until he was about six months old and very active. Then she found a babysitter, and to her surprise,

---

[3] Elizabeth Holmes is set to be sentenced on September 26.—*The New York Times* (nytimes.com) viewed January 30, 2022.

many people including the librarians complained that they rarely got to see him anymore!

About nine months later, Christine and Michael Economides, who were graduate students in the Petroleum Engineering Department, which shared a building with the Geophysics Department, had a little boy. Their baby (another Alexander) moved into the building in a baby carriage. The staff considered Eve's baby to be the "Geophysics Baby" and Christine and Michael's to be the "Petroleum Engineering Baby." Eve didn't get to know Christine and Michael until after their baby was born, but they became lifelong friends with valuable business connections.

Part III

Career Challenges

"Such bees! Bilbo had never seen anything like them.
If one were to sting me," He thought "I should swell up as big as I am!"

—J. R. R. Tolkien, The Hobbit

7

Bullying

When we talked to people about the problems they encountered with mentoring and sponsoring, many of them described experiences which we recognized as bullying. Bullying is deliberate and repeated misuse of real or perceived power. Sponsors are in a position of influence and power, and mentors may learn information from their confidential discussions that they choose to use to the detriment of their mentee. Both sponsors and mentors can morph into bullies.

Bullies attack their victims verbally and/or physically inflicting psychological and/or physical damage. If the victim is lucky, their encounter with the bully is short. When the bullying continues over an extended period, the mental, physical and emotional damage is greater and the recovery period longer. The victim's professional reputation may also suffer from the slander the bully promulgates about them.

Recovery from mental battering alone may take a long time. Repairing your personal and professional reputation also requires time and effort. The sooner you recognize and "label" or categorize your problem, the more quickly you can begin your recovery.

7.1 Madhuri's Problems

When Madhuri shared the problem she had, we realized her manager was bullying her. At her company, a female manager who was about fifteen years

© The Author(s), under exclusive license to Springer Nature
Switzerland AG 2022
E. Sprunt and M. A. Capello, *A Guide to Career Resilience*,
https://doi.org/10.1007/978-3-031-05588-1_7

older was officially her mentor and responsible for her performance review. Madhuri tried for months to please her boss, but her boss always labeled her work as unsatisfactory. The boss's requests, which were given verbally, changed daily. Mostly, she asked Madhuri to perform "busy" work, which was not linked to the long-term success of the company, and never provided growth opportunities. Madhuri's physical and mental health suffered from the bullying. She felt depressed, was sick more often and worried about what would happen to her career. The company did not have an ombudsman program, so Madhuri began looking for growth opportunities outside of her organization and a mentor who would not communicate with her boss.

Her parents, who were living far away in another part of India, became a critical source of support. Through phone calls, she kept them informed about what was happening. They said she was learning the tricks of a terrible boss and the experience would strengthen her. Madhuri credited their advice with helping and enabling her to focus on the long term and learn to be stronger and more resilient. She was lucky to have their emotional support, but her parents' advice encouraged her to be passive and submit to the abuse.

Madhuri suffered from her abusive boss's behavior until one of her colleagues spoke to the head of the business unit. Her colleague's intervention prompted the company to transfer the bully. Madhuri said that, to her knowledge, they took no other actions to penalize her former boss.

Madhuri's parents' advice (and perhaps her culture) encouraged her to submit and endure the abuse. Other families or cultures might take a more confrontational approach. A culture of submission enables prolonged and highly destructive bullying. When bystanders witness and ignore inappropriate behavior, it empowers bullies to continue their misdeeds.

7.2   Label the Behavior as Bullying

To reduce the incidence of bullying, organizations should encourage a culture that, "*If you see bullying, say something.*" The most effective way to deal with bullying is to halt the bullying behavior before serious mental damage occurs. As soon as you recognize behavior as bullying, call the behavior by its name "bullying" and act.

As per the American Psychological Association, bullying is a form of aggressive behavior in which someone intentionally and repeatedly causes another person injury or discomfort. Bullying can take the form of physical contact, words or more subtle actions.[1]

They specifically state that *"The bullied individual typically has trouble defending him or herself and does nothing to 'cause' the bullying."*[2]

Bullying is ever present in society, and over time has taken many different forms. Currently, cyber-bullying is rampant, especially on social media, as described by Forbes Magazine on September 29, 2021.[3]

All too often, witnesses feel that their best protection from the bully is to stay silent. We need to encourage cultures to have zero tolerance for bullying. Employers need to make witnesses feel they can support their bullied colleagues without being penalized.

Find help inside and/or outside your organization as soon as possible. Use the word "bullying". Silence and submission are not the solution. Bullying is an ongoing and deliberate misuse of power in relationships through repeated verbal, physical and/or social behavior that intends to cause physical, social and/or psychological harm. Document the situation and file a formal complaint with your management, even if you must skip up a few levels in your management ladder.

If you feel you can't complain to your management, immediately begin looking for alternative employment. The time to look for a job is before the bully erodes your self-confidence. Break the cycle before it breaks you. The new job may be in another part of your current organization. Eve did that when she had a boss who predicted a year in advance that she would not have a good performance review, because following a downsizing reorganization in which half of her prior group was terminated, she was transferred into a technical area in which she had no prior experience.

Sadly, there are many bullies in the world. After suffering from bullying, watch for signs of a new bully and take appropriate action. Support others whom you see being bullied. Especially if you see bullying occurring in a mentoring relationship.

Talking with Madhuri brought back terrible memories for Eve of being bullied by Peggy, who was Eve's boss in the late 1980s. Peggy was the first female Ph.D. in the company. She was hired in the 1950s after a long and nearly futile job search. When Eve joined the company a couple of decades later, Peggy reached out to mentor Eve. Peggy often shared stories about

---

her impoverished childhood and the tremendous obstacles she had to over-come to find employment. She explained she never had children, because her husband had a heart condition and expected to die young, and he did. Peggy believed if she became a widow, she wouldn't be able to provide financial support and care for children.

When Eve joined the company, she had a Ph.D. and a child. A few years later, she had a second child. It wasn't easy working full time with two young children, but she managed. Before Peggy became her boss, Eve was able to decide for herself which parts of Peggy's advice she implemented, and which she ignored. Her career was going well, because in her first seven years Eve had two major technical accomplishments in which she identified why petroleum reservoirs were not behaving as expected and received four promotions. Then Peggy became Eve's third boss at that company.

Eve's life turned into a nightmare the day Peggy walked into her office, looked at the pictures of her children and said, *"You've had everything!"* It was a declaration of war. Peggy tried to retroactively reassign credit for Eve's major accomplishments to other people and did everything she could to destroy Eve's reputation as a top performer.

Eve had always had high marks for her communication skills on her annual assessment until Peggy became her boss. Peggy graded her writing skills as inadequate the same year the Society of Petroleum Engineers asked Eve to serve as Executive Editor for one of their technical journals. Despite ample proof that the ratings were discriminatory and inappropriate, Eve could not get them changed.

When Eve tried to transfer to another position in the company within commuting range, Peggy sabotaged her efforts. Peggy made Eve's life at work a living hell. The constant bullying undercut Eve's emotional equilibrium, making her emotionally fragile.

Although the job market was poor because of falling oil prices, Eve sought other employment. When a pair of interviewers from a competitor probed Eve about why she wanted to switch companies, Eve teared up, ruining her chances of securing a job offer. It is much harder to shine in a job interview when you are emotionally fragile from intense bullying.

Eve's recovery began when she realized Peggy was bullying her and began reading about bullies in the workplace.[4] During a break in an offsite company course on soft skills, Eve shared her problems with the instructor, who was a consultant with influence at corporate headquarters. Eve explained to the instructor how Peggy was bullying her. A few weeks after the course, Peggy

---

[4] https://workplacebullying.org/about-us/.

called Eve into her office and said that the instructor from the course would tutor her on how to improve her soft skills.

When the instructor visited Eve, he said, "*Be patient. Your situation will improve within six months.*" While the prospect of six more months of hell was not appealing, it was better than an indefinite sentence or loss of her job. Shortly after meeting with the instructor, Eve had surgery and took three weeks of medical leave. During Eve's absence, Peggy announced she was retiring and departed. Not only was Eve spared seeing Peggy honored in retirement festivities, but by the time she returned from her sick leave, Peggy was gone.

Within the company, it took several years for Eve to restore her reputation. At first, her new boss kept her on a very short leash. When he retired a few years later, he called Eve into his office and explained that Peggy had filled Eve's personnel file with very damaging material, which he had removed and shredded.

7.3   Stop the Bullying ASAP

Eve thought after her experiences with Peggy she was bully proof, but she wasn't. About fifteen years later, at another company, a man, Aarush, who had groomed Eve by mentoring her, tried to bully her. During the grooming phase, the man, a tall, dark Indian, explained how in India he faced discrimination because of his dark coloring, and described how his father, a high-ranking military officer, arranged his office to intimidate people. As soon as Eve asked to switch onto the team with Aarush, he morphed from being a mentor into a monster. When Eve realized she was so intimidated that she was afraid to stay late in the office without witnesses, she knew it was time to blow the whistle on the bully.

Unfortunately, when an employee seeks protection from another employee's bullying, it becomes a "*he said, she said*" situation. Aarush was a favorite of their mutual boss, but Eve's sponsor was her boss's boss. Eve's attempt to break out of the bullying cycle turned into a traumatic witch trial, with Eve accused of being the witch. The company decided both Eve and Aarush should take remedial training. Eve immediately did. But Aarush may never have completed the training.

Eve escaped from Aarush's bullying before incurring deep psychological damage, but she feared he had damaged her reputation. Fortunately, the duration of the bullying was brief, and the emotional and reputational damage was not as serious or long-lasting as that caused by Peggy. Eve emerged, ready for her next challenge. If you quickly recognize that you are being bullied before you get caught in a downward spiral and escape from your tormentor, you can minimize the emotional and reputational impact and recover faster.

7.4   Long Lasting Impact—Persistent Psychological Damage

Another of the people Eve and Maria interviewed, Belle, shared her story of long-term wounds inflicted by a woman, who she had considered to be a mentor and a role model. When we spoke with her almost two decades after the bullying episode, Belle was still reeling from her experiences. She said, "*It was a very troubling experience that came from a woman that I looked up to. Afterwards, I no longer believed in rainbows, especially coming from women. I have become a very guarded individual in terms of trust and looking up to people.*" In Belle's case, the female bully was about twenty years older than her victim. Belle thought her female boss felt so threatened by her ability that everything she suggested, her boss seized as an opportunity to intimidate her.

When Belle couldn't take it anymore, and thought she would have to quit, she reached out for help to the man at the top of her organization. She said, "*I was just trying to do the best of my ability. To get this shit (sic.) for no reason whatever. Back then, I didn't know the game. I didn't know politics and a lot of things I now understand. I was going into this sea with a very clean soul.*" The top man was a very proper noble person, who supported Belle in her time of need and continued to support her so much that people rumored she was his niece.

Because of the bullying Belle experienced, she said, "*I do not socialize at work. I never mix personal life with professional activities. My work is not my life. This determination comes from being kicked very badly. Until I was about 33, I continued to be a believer in good. I would always give people the benefit of the doubt. Now, before I trust anyone, they have to prove they can be trusted.*"

She added, "*All the bad things that have happened have made me pretty much know myself… Being nice requires a mask. I'm a chameleon. I use a lot of masks… I'm self-sufficient. Only when I'm in an immense dilemma, will I go out and ask for advice. I've created a sort of box within a box within a box with the last box… for me.*"

Talking to Belle showed how brutal bullying by a former role model can change a person's entire outlook and life.

7.5   Maria Angela's Experiences Overcoming Bullying

Maria Angela experienced severe bullying when she was in elementary and high school. She was different—an Italian girl in a Venezuelan school. Her top grades and shyness made her a target for other students who were learning to be bullies. Children can be cruel. Episodes of being laughed at for her glasses adversely affected her self-esteem for years. Bullying destroys the victim's self-esteem. Bullies get a thrill from intimidating talented and previously confident people.

Within her first two years of starting work at Lagoven, the former Esso Venezuela, Maria Angela was targeted by several co-workers including Enrique, who was very harmful, because he was admired for earning a Ph.D. in the United States and expected to be quickly promoted to be a Team Leader. Enrique always mocked Maria Angela, both in private and in front of colleagues. The worst of his attacks was at the farewell dinner of a field trip. Enrique deviated from the planned program and presented each participant with his own unsolicited awards. He presented every member of the team with a nice award such as *"Best Fossil Finder," "Top Mountain Crawler,"* or *"Best Swimmer"* until he got to Maria Angela. After praising everyone else, he gave her the *"Golden Lemon"* explaining it was awarded for her acidity and harsh communication style.

Maria Angela was astonished, but her self-healing began when she noticed no one applauded. The group demanded Enrique withdraw that award and behave professionally. When she returned to the office, many of her co-workers, including some she didn't know, spoke to her. They wanted to explain that they had observed Enrique picking on her and making jokes at her expense, but they had done nothing to help her. They recognized they could have stopped his misbehavior that they considered "picking" on someone. It was bullying. Individuals and organizations should not try to downplay inappropriate behavior by using softer descriptions. Labeling the misbehavior as bullying is the first step to eliminating it.

The manager asked both Enrique and Maria Angela to a meeting. He wanted Enrique to apologize to Maria Angela, but Enrique did not attend. The organization's tolerance of bullying was further demonstrated when, soon after this episode, Enrique was promoted to be a Team Leader.

7.6   Seek Help Immediately

Maria Angela learned you must trust your instincts. If you do not feel comfortable with your workplace, something is not right. If you are feeling bullied, seek help immediately. Recognize the signs: repeated impossible requests, attention focused on you and you only, damage to your reputation, and being ridiculed in front of others or in private. When nothing you do is good enough to satisfy the bully and you feel diminished by the bully's feedback, you need to find a way to escape the situation.

Listen to your co-workers. They may recognize a bullying situation in which you are so immersed, your judgement is impaired. The sooner you break the cycle of bullying, the more likely you are to fully recover.

When you see bullying, reach out to help the victim. Even if you are afraid to speak to management, you can provide emotional support and guidance to them. If you are a silent bystander, you are part of the problem.

8

Sexual Predators in the Workplace

Close one-on-one interactions with mentors and sponsors are ripe with opportunities for abuse. Often, these private conversations occur in locations without witnesses. The mentor or sponsor, who is in a position of power, may seize the opportunity and morph into a sexual predator.

Many people take advantage of their position for inappropriate reasons. Both mentors and sponsors can become sexual predators. Inexperienced young people may be flattered when an important individual takes a special interest in them. They need to know that the best time to get help is the moment they realize there is a problem.

8.1 Problems Start at a Young Age

Mentors can also be sexual predators. There always seems to be an abundance of stories about religious clerics who misuse their position to molest children whose parents expected them to provide spiritual training, but they are not the only large group to fall from grace. Many Boy Scout leaders also abused their positions to assault the young boys who they were supposed to mentor. In a July 1, 2021, headline, NBC News wrote, "*Tens of thousands of people who say they were sexually abused while scouts and filed suit against the Boy Scouts of*

E. Sprunt and M. A. Capello, *A Guide to Career Resilience*, https://doi.org/10.1007/978-3-031-05588-1_8

America have reached an $850 million settlement, the largest in a child sexual abuse case in United States history."[1]

As a parent, you can teach even very young children about behaviors they should report immediately. When Eve was five, an older boy in her neighborhood got her alone, pulled down her pants and exhibited himself. Eve immediately ran screaming to her mother. The boy begged Eve's mother not to talk to his mother, but she did. It was a powerful lesson to Eve that she didn't have to face problems like that alone.

Keeping children in ignorance of inappropriate behavior makes them more vulnerable. If a child learns at a young age that they should report figures of authority who touch them inappropriately, they will be better able to protect themselves when they become a teenager and later a young adult.

8.2   Predator Warning Signs

When Eve was a freshman at MIT, she thought she wanted to be a physicist and enrolled in a seminar with a junior physics professor. One day, a few weeks into her first term at MIT, she received a call from Professor T, explaining that he was cancelling the seminar that day, but could meet with her to discuss the material. Eve naively said, "*Yes.*"

Prof. T. showed up outside her dorm in a VW bug and drove her out to Walden Pond. Eve spent the afternoon saying, "*Yes, Prof. T. No Prof. T.*" After Eve repeatedly refused to lie down on the grass in a secluded location, Prof. T. drove her back to her dorm. Later that year, Eve learned one of the other 72 female freshmen was "having an affair" with Prof. T. He became a well-known physicist, but Eve never heard about him being censured for inappropriate involvement with female students.

Sexual predators recognize and leverage their power and influence on their prey. However, you have much more power than you think. Eve knew that if Prof. T. abandoned her at Walden Pond, she would have trouble getting back to MIT, but she maintained control by insisting on staying in a standing position and addressing him decisively and formally.

You need to know your boundaries and what is important to you. Sexual predators use their power and influence to manipulate their prey, but you have more power than you recognize. Prof. T. caught Eve off guard and whisked her away to a location in which she was vulnerable to his overtures. Once he had her in his control in his car, Eve didn't respond to any

---

[1] https://www.nbcnews.com/news/us-news/boy-scouts-reach-850-million-settlement-tens-thousands-sexual-abuse-n1272955 as of December 13, 2021.

of his overtures. When someone makes inappropriate advances, stop them long before you get to your limits. Eve did and continued as a participant in the physics seminar without further incidents or problems.

Maria Angela's approach to advances is to always say, "*No!*" It does not matter if it looks like a simple invitation for coffee or lunch. "*No!*" is your first and strongest line of defense. If a person at work invites you and you alone to go somewhere, something is not right. Try "to see it coming" and discourage them. If it happens, just say no. People's intentions of romantic involvement are often noticeable. Avoid romantic overtures by leaving meetings as soon as they end, and not arriving too early. If you work alone with just one other person, always leave the door open. If you must work with a person you distrust, take frequent breaks to go for water or to call someone, even if those breaks are not needed. Despite taking these precautions, people perceive Maria Angela to be very social and open. Do not be afraid to say, "*No!*" Don't mix business and pleasure.

If you are romantically involved with a person above you in the hierarchy, investigate that person's reputation. There is often considerable chatter about the behavior of individuals who abuse their position for sexual favors. If there is, don't fool yourself that you are special. If you are certain the relationship is sincere, consider notifying Human Resources, but be prepared for negative surprises. When you become romantically involved with someone above you in your organization's hierarchy, you are putting your career and your entire future at risk.

8.3   A Long Standing Issue: Sexual Aggression

If you are new to an organization, you may be struggling to build a network and advance. You need clear boundaries for all your interactions with potential mentors and sponsors. Sponsors, even more than mentors, can leverage their gatekeeper powers to secure sexual favors. There are many examples of those who can influence the staffing of highly desirable positions, using their power in inappropriate ways.

Eve discovered in her mother's diaries an example of a mentor/sponsor who first groomed and then abused young women. In 1940, when Eve's then twenty-year-old mother, Ruth, was trying to find her first job in Washington D. C. as a commercial artist, she was referred to Mr. Janof at the Washington Post. She described him as "*very handsome and very nice to me.*"

For about three months, Ruth had tried without success to find a job. When she met Mr. Janof, he gave her some ads to copy, called a local store, and lied telling them she had worked for him. He arranged an appointment for Ruth to meet the hiring manager. While Ruth didn't realize it, the store knew Mr. Janof was lying, but he had enough influence to get her part-time work with irregular hours. When she returned to see Mr. Janof, he suggested she "pad" her bill. Ruth's reaction was, "*The crook—I don't trust him.*"

Ruth didn't trust Mr. Janof, but she knew he could get her employment. When Ruth wanted to move to a better job, she went back to see Mr. Janof, who recommended her to another store, where she got a full-time job, which required her to move away from home. During her thank you visit to Mr. Janof, Ruth mentioned she had stayed at the "Y," which was a safe, secured, chaperoned place with rooms for women. Ruth described her encounter. "*He said I was the kind of girl who'd stay at the Y—in between pinching my cheek and chucking me under the chin.*" This was the first time Mr. Janof touched Ruth, who, although she was now 21, was a naïve young woman who had never been kissed.

Two years later, when Ruth again needed help to find a job, she returned to see Mr. Janof. This time, things got more physical. Ruth recorded, "*Janof came in and hugged me. I enjoyed this tremendously. He was busy, so I didn't really get to talk to him, but hugging him was fun.*" Mr. Janof was slowly grooming Ruth to be abused. After her next visit, she noted, "*he told me about his son Donald... managed to adroitly kiss my forehead and cheek. I enjoyed it.*"

As soon as Janof began discussing getting Ruth a job at the Washington Post, his aggression rapidly escalated. "*Janof took me into his office and told me he was working the pros and cons of a job for me at the Post. Janof keeps grabbing me, but I won't let him kiss me on the mouth.*"

Having groomed Ruth for two years, Mr. Janof was ready to pounce. Ruth wrote, "*Worked at the Post again this evening. Janof came in and grabbed me. Golly, he's rough.*" A week later when Ruth went to see Janof after work, "*He pinched and beat me, kissed my cheeks and forehead, but didn't kiss my mouth—I didn't exactly respond—much as I wanted to.*" The next time she saw Janof, "*He lectured me on morality, meanwhile pinching me and kissing my neck.*" At first, Ruth enjoyed it. "*Janof grabbed me twice. Once at my drawing board and once in his art studio. I loved it.*" After a month working at the Post, Ruth began to find Mr. Janof's behavior offensive, and described him as "*kissing me in a perverted sort of way. I feel soiled somehow.*" When she told a male colleague that his portrait of Janof lacked the horns, "*He relayed the message to Jan, who laughed.*"

Over the following months, Janof's behavior escalated.

Janof came over and grabbed me, licking the back of my neck.

Had lunch with Marion, who told me how Janof mauled her yesterday. When we were in the art room, he spoke to her and ignored me.

After work when I was waiting for the elevator, he—under the pretext of wanting to tell me something, kissed my cheek.

At noon, I had a slight encounter with Janof in the Art room. He's so unexpected. He grabbed me and shoved his leg between mine, while demurely kissing me on the brow.

The abusive work environment was widespread in Washington at that time. When Ruth, as part of her job at the Post, had to go to a store to sketch, she wrote, the store owner kept "*finding various occasions to put his hand on my knee.*" That store owner soon escalated to trying to kiss her, so Ruth tried unsuccessfully to arrange a chaperone. Her refusal to be alone with the store owner was a factor in the loss of her job with the Post.

Ruth's refusal to acquiesce to Janof's overtures didn't help. "*This evening in the art room, Janof grabbed me, and I drove my fingernails into his hand. He knocked me over, catching me before I hit the floor.*"

But other men joined the assault. "*Patrick came up and started licking the back of my neck. He was delighted because I began shivering.*"

"*Patrick came up to me and whispered, 'Someday I'm going to get you in the corner and rape the Hell out of you—and you'll love it.' Later he planted a juicy kiss on my cheek.*"

When Ruth decided to move to New York to find work, Patrick said her samples could be shown anywhere and gave her the name of a man who had an agency in New York. "*Says he's not a wolf!*" And the man in New York who Patrick told her to contact wasn't. In New York, which had far more opportunities for young female artists, the harassment was not nearly as bad as in Washington D.C., which had very limited opportunities for artists. Harassers are everywhere, but the more opportunities a person has, the less likely they are to suffer from unwelcome advances.

Ruth was vulnerable to harassment from a sponsor, because in the early 1940s, before American men were called up to serve in the military, it was very difficult for women to get employment. A strong sponsor made all the difference in who got hired and who didn't. Even if a woman was hired, it was very difficult for her to get a raise or a promotion, because the bias was to give better positions to a man. This created a situation in which a powerful, highly networked man like Mr. Janof could bend women to his will.

8.4   An Oil Field Experience

When "Gabriella" began working in the oil field in Texas, they gave her a couple of mentors. One was the only other woman at that location, who went from "*sweet to nasty within a month.*" The other was a seemingly "*nice, older gentleman.*" As the only female engineer at her work location, Gabriella faced intense criticism. "*Everything was a put down—my accent, my hair.*" Her male mentor taught her everything. He was always very kind until after she returned from maternity leave.

Gabriella was still breastfeeding her baby when she returned to work and her breasts were "*BIG*". Her male mentor, an old, white-haired guy, who had been working for forty something years, started making dirty jokes. Gabriella didn't immediately complain, because "*He was the only one who was nice to me.*" Her mentor's behavior got more and more offensive.

She said, "*I would tell him to stop it and thought I could handle it. He was nice and would teach me and was the only engineer who would have my back. I thought I could not go to anyone else, but him and thought I had to listen to learn. When we were riding in his car to the work site, I had to listen to his horrible details about his sexual life. I buttoned my shirt up so nothing was showing. When I bent over in the car, he pulled on my pants, tried to touch my butt, and made a nasty comment. He knew I was bothered because I was silent and I'm not usually silent.*" "*I was shocked. Everyone knew he was a dirty man. It was not a secret. Everyone called him 'Dirty Randy.' I thought I could handle 'Dirty Randy,' because he was my mentor. I thought I was under control. I had a cell phone, but I didn't even think to use it.*"

"*One day I thought, I'm gonna talk and say what is happening, but I knew I could not go to the lady (her other mentor, who was a Queen Bee). The secretaries were always very kind. They knew about Randy and said it was his normal behavior. Maybe they thought it was only verbal. I spoke to the young male boss. He was upset and said I should never have to go through that, apologized and called HR* [Human Resources]. *Randy had texted me things and I had his voicemails, so I had all the proof. HR said, 'Well, we will make sure it doesn't happen.'*"

When Gabriella spoke to the Queen Bee, she said, "*We all know about Randy.*" She didn't want to cause drama and provided no support.

Gabriella said, "*Secrecy protects the guilty people.*" and went to speak to a male director, who asked, "*What do you want me to do?*" She said, "*You have to get rid of this guy. I don't want another girl to have to put up with this treatment. The response of that director was that he thought I was a troublemaker and said he did not believe me. I asked myself, 'Why would I put myself in this story? What* do I have to gain?'"."We went back and forth. They put a knot in the system, but I said they had to make Randy accountable, so they offered him a retirement package and he took it. They were going to have a party for him, but I said I would stand up and speak in front of the entire company and said he needed to be gone with shame without a party. They did not have a company party for him, but they had a good luck party for him in a restaurant with a few people."

"I spoke to my boss before my husband. When I finally told my family, my husband could not understand how I allowed that behavior for so long. It was hard for him to get it and it turned me upside down. It struck the core of me and how I worked so hard to be where I am. When I'm finally there and find that I must continue paying. That made me so strong. Now, when I see it, I call it right away. I tell men to keep their hands to themselves and stop it immediately. I will not put up with any of that type of behavior physically or mentally. We don't have to put up with that to gain reputation, just because we don't have experience. It is not the price anyone should pay to have dignity. I thought it was the price I had to pay."

"It took years to recover. I help young engineers and tell them, 'You don't have to sacrifice your body or your mind to learn.' I have healed myself by helping others."

8.5   The Entertainment Industry

The entertainment industry is especially infamous for men in power taking advantage of their positions. The phrase "casting couch" refers to actors or actresses being rewarded with parts in films, plays, or other productions in return for providing sexual favors to the director or other individuals with control over casting. Harvey Weinstein[2] is a particularly notorious example, but far from the only one. The famous director, Alfred Hitchcock, had a terrible reputation.

For decades before Harvey Weinstein was convicted and jailed in 2020 for rape in the third degree and a criminal sexual act and sentenced to 23 years of imprisonment, he was an incredibly successful film producer. His blockbuster films won prestigious awards and earned enormous amounts of money. All the while, rumors of his sexual aggression were rampant. In 1998, Gwyneth Paltrow, who starred in Weinstein's Academy Award-winning film, ***Shakespeare in Love***, said on David Letterman's Late-Night Show, Weinstein "*will coerce you to do a thing or two.*" The warnings kept coming, but Weinstein

---

[2] https://en.wikipedia.org/wiki/Harvey_Weinstein as of December 21, 2021.

seemed immune to legal action until October 2017, when two female New York Times reporters, Jodi Kanto and Megan Twohey, wrote a story accusing Weinstein[3] of for three decades sexually harassing actresses, female production assistants and others. Wikipedia lists over one hundred women who have accused him.[4]

Comedian and actor Bill Cosby[5,6] had a reputation as a nice guy and the nickname "America's Dad" until 2014, when a young Black comedian, Hannibal Buress, encouraged people to do a Google search on "Bill Cosby rape." Cosby's nice guy image in the highly successful Cosby Show as the mentoring dad, had protected him for decades. That image began crumbling when Barbara Bowman of Arizona described how, in late 1985 or early 1986, about the time she turned eighteen, her agent introduced her to Bill Cosby in Denver. Barbara had worked with that agent since she was fourteen and was thrilled her agent, *"thought enough of me to introduce me to Bill Cosby. He was going to groom and mentor me. I was going to New York."*[7] Her first meeting with Mr. Cosby was very professional, so she moved to New York.

In New York, Cosby invited Barbara to his home for dinner. After one glass of wine, she felt dizzy and sick, and then passed out. She thought she had just gotten sick. For their next encounter, Mr. Cosby invited her to Atlantic City. When Cosby threw her on his bed and attempted to undress her and himself, she made so much noise; he threw her out. Then her agent abandoned her, and she moved home to Denver.[8]

Throughout the decades when he was revered as America's Dad, Bill Cosby drugged and violated aspiring young women, who he chemically rendered unable to defend themselves. Over thirty women accused Cosby of misconduct in incidents as early as the mid-1960s. Finally, in 2018, he was convicted of three counts of indecent assault against a woman and sentenced to three to ten years in prison. The unmasking of Bill Cosby as a perverted serial sexual aggressor came as a huge surprise to his many fans.

The Bill Cosby case was shocking. It shattered widespread public perception that sexual predators were easy to identify. Since then, multiple other cases have filled the news.

---

3 https://www.nytimes.com/2017/12/07/books/jodi-kantor-megan-twohey-book.html.

4 https://en.wikipedia.org/wiki/Harvey_Weinstein as of December 21, 2021.

5 https://www.bbc.com/news/entertainment-arts-30194819#:~:text=Bill%20Cosby%20was%20once%20known%20to%20millions%20as,prison%20after%20serving%20two%20years%20of%20the%20sentence, as of December 12, 2021.

6 https://en.wikipedia.org/wiki/Bill_Cosby, as of December 12, 2021.

7 https://www.phillymag.com/news/2006/11/01/cosby-threw-me-on-the-bed/.

8 ibid.

Barbara Bowman's comments about her expectations that Cosby would groom her for success as an actress illustrate how mentors and sponsors abuse those who seek their help. Acting is a notoriously tough profession to break into, but it isn't unique. Getting started in many professions is challenging. Mentors and sponsors in all fields have ample opportunities to prey on those who seek their assistance.

8.6   Well Known Secrets

There has long been chatter about some women "sleeping their way" to the top. This was especially true when there were very few female executives. A notorious example was Mary Cunningham, who was born the same year as Eve. After graduating from Harvard Business School, Ms. Cunningham accepted a management position as executive assistant to the CEO of the Bendix Corporation, William Agee. Cunningham was quickly promoted to Vice President of Strategic Planning by Agee. Less than two years later, after public accusations of her affair with Agee, Cunningham resigned from Bendix. A couple of years later, after divorcing their spouses, Agee and Cunningham married. Stanford University Business School used Cunningham's experience as a case study in its course, "Power and Politics in Organizations."[9]

Maria Angela learned early in her career that not all people in power positions were well-intentioned. Stories were rampant in Venezuela about sexual favors being exchanged for positions or promotions. She witnessed sponsoring loops in which sponsors requested sexual favors in exchange for a promotion or an endorsement. She also observed individuals offering sexual favors to bosses or leaders to advance more rapidly. It was not the norm, but it was present. She did not observe this in the Middle East, where social norms are much stricter.

Early in her career as a geophysicist, Maria Angela had to work with geophysical acquisition crews at drilling sites. She was the first woman to work in the field in Venezuela and possibly in all Latin America. The first time Maria Angela worked in a seismic recording truck, she found its walls covered with posters of naked women and other more explicit pornographic posters. The situation inside the truck was not comfortable for her. When she returned to the office, Maria Angela told her manager about it, and he said, *"Tomorrow, I will go with you to the field."*

---

[9] https://alchetron.com/Mary-Cunningham-Agee.

The next day, inside the cabin of the recording truck, the manager greeted everyone, and then in a very friendly way said, "*Listen guys, Engineer Capello will be coming here, and I think we all agree we cannot have things here you would not post in your own kitchen at home for your wife and children to see.*" That solved the problem. Together, everyone, including Maria Angela, pulled the posters off the walls and the issue was forgotten. As the first woman in Venezuela to serve as an oil field supervisor, Maria Angela was pleased that men in operations were extremely respectful to her.

8.7 Men Are also Targets

It is not just women who suffer from unwelcome sexual overtures. An article in the New York Times on January 31, 2021, describes how, "John Weaver, a longtime Republican strategist and co-founder of the prominent anti-Trump group the Lincoln project, has for years sent unsolicited and sexually provocative messages online to young men, often while suggesting he could help them get work in politics, according to interviews with 21 men who received them."[10]

Male or female, if you are seeking a career in a field in which sponsors are critical, you will be vulnerable to inappropriate behavior. Never let a sponsor push you into doing something you do not want to do. You always have more control and options than you think you do.

8.8 Everyone's Problem

Sexual abuse associated with mentoring and sponsoring includes both sides: the powerful and experienced person and the junior person in the relationship. Often within the organization, many people are aware of the situation, but it is not "officially known." Whispered, but unverified relationships were private matters, not organizational issues. However, private relationships in organizations are everyone's problem. In the absence of whistleblowers, these liaisons create a toxic environment ripe for unfair hiring and promotion practices and sexual harassment.

Leaders should protect the members of their communities by setting the right expectations, whether it is in an educational environment or in a work location. Everyone should be able to study or work without harassment.

---

[10] https://www.nytimes.com/2021/01/31/us/politics/john-weaver-lincoln-project-harassment.html.

Organizations can help prevent sexual harassment by providing ombudsman, whistleblower programs, and clear rules on prohibited behaviors. These programs and rules can nip problems in the bud and minimize damage.

If you cannot find help within your organization, contact professional societies for help. Many volunteer organizations have specialized groups that can provide advice if you are encountering a lack of support and problems in your own organization. Don't suffer in silence. Find help when you need it.

9

Queen Bees

9.1　Insect and Human Behavior

The Merriam-Webster dictionary has two definitions for "queen bee."[1]

*1:	the fertile fully developed female of a social bee (as the honeybee)*
*2:	a woman who dominates or leads a group (as in a social activity).*

Beehives only have a single queen at any given time. Hives hatch multiple queen bees when they need a replacement queen. The newly hatched queen bees are notorious for battling each other until only one remains alive. Male or drone bees' sole function is to mate once with the new queen. Worker bees in a hive are female, but do not reproduce.

Three psychologists at the University of Michigan, Graham Staines, Carol Tavris, and Toby Epstein Jayaratne are credited with being the first to draw the analogy between dominant women in the workplace and queen bees in their article published in 1974.[2] Both Eve and Maria Angela have suffered from working with female executives, who were known to their staff as Queen Bees (Fig. 9.1).

---

[1] https://www.merriam-webster.com/dictionary/queen%20bee.
[2] Staines G, Tavris C, Jayaratne TE (1974). The queen bee syndrome. Psychol. Today 7:63-66.

© The Author(s), under exclusive license to Springer Nature Switzerland AG 2022
E. Sprunt and M. A. Capello, *A Guide to Career Resilience*,
https://doi.org/10.1007/978-3-031-05588-1_9

Fig. 9.1 The photo shows a Queen Bee surrounded by her court, in the apiary of Margarita Alberdi de Genolet, located in Kassel, Germany. Every year, the queen bees are marked with color dots, to facilitate their differentiation, and in 2021, the selected color was white. The hive needs a queen to survive. The female worker bees that surround her are called the "court", and they take care of cleaning the Queen, and feed her

9.2   Male Dominated Work Environments

Women were almost non-existent in management in many organizations in the early 1970s. In the United States, the Equal Pay Act of 1963[3] created a labor standard that required employers to pay men and women the same wages when they performed jobs that were equal or essentially the same. The Civil Rights Act of 1964[4] prohibited discrimination based on sex, race, color, national origin, and religion. Despite that legislation, when Eve was hunting for a summer job in 1969, the "help wanted" advertisements in the newspapers were divided into "male" and "female" sections. The opportunities for women were poorly paid, dead-end positions like receptionist, keypunch operator (to type the cards for the computers that only took input through

---

[3] https://www.eeoc.gov/statutes/equal-pay-act-1963.

[4] https://www.dol.gov/agencies/oasam/civil-rights-center/statutes/civil-rights-act-of-1964.

punched cards) and secretaries. Eve could not find a job for the summer even though she had been accepted to MIT. Finally, her father leveraged his network and got her a position as a clerk in a bank. Her major accomplishment that summer was finding a discrepancy that led to a male bank teller being caught embezzling funds to go to the Woodstock Music Festival.[5] The teller was fired for *"carelessness with the bank's records"*.

After examining promotion rates and the impact of the women's movement on the workplace, Staines, Tavris and Jayaratne (1974)[6] concluded that women who achieved success in male-dominated environments were more likely to oppose the rise of other women. Their thinking was that the patriarchal culture of work encouraged the few women who rose to the top to become obsessed with maintaining their authority.

In hindsight, it is amazing that Staines[7] and his co-authors were studying dominant women in the workplace in the early 1970s. When Eve took her first permanent corporate position in the late 1970s, there were only four other female technical professionals in the large research laboratory. One was hired the previous year, and another was a male employee's spouse. When the lab was created a couple of decades earlier, women were completely banned and even the secretaries were male. One male secretary was still working when Eve began working there. Eve became friends with one of the female librarians, who revealed she had been required to resign every time she became visibly pregnant and was rehired after the birth of each of her three children. The company still did not provide any maternity leave, but at least did not fire Eve when she became pregnant in 1980.

9.3   Token Women

When women began to gain entry to professional roles, they often did not have an equal opportunity to advance. The few women who moved into management roles tended to perceive every other competent woman as competition. Just as in the beehive when multiple infant queen bees hatched, the competition for the crown was brutal. When there was no chance of advancement, the women were more collaborative and mutually supportive.

---

[5] https://en.wikipedia.org/wiki/Woodstock#:~:text=Woodstock%20Music%20and%20Art%20Fair%2C%20commonly%20referred%20to,%2865%20km%29%20southwest%20of%20the%20town%20of%20Woodstock. Viewed 2/1/2022.

[6] ibid.

[7] Staines G, Tavris C, Jayaratne TE (1974). The queen bee syndrome. Psychol. Today 7:63–66.

When a woman was selected as the token female leader, she feared all the other talented women as competition for the few token positions.

You can benefit by watching human Queen Bees and analyzing what they have done to become so powerful in an organization. Like their insect name-sakes, they perceive other females as competitors who must be eliminated. Your organization's gossip "grapevine" may be useful for learning whether bringing yourself to the attention of certain female leaders is a career limiting move. They may pretend to mentor you, only to stab you in the back. With more and more women moving up to high positions in organizations, hope-fully queen bees will perceive other women as less of a threat and lose some of their sting. so their wounds are no nastier than those from the average bad boss.

9.4   Head of the Women's Network

Eve worked in a corporation in which the head of the company's women's network was an attractive and superficially friendly woman who loved to talk about her career. This Queen Bee was sought after as a speaker at women's events. She would explain how her husband quit working to provide support to her in the same way other top executives' wives did for their husbands. Her husband would tell people at social events that his job was to support the Queen Bee. The problem was the Queen Bee didn't think it was possible to have a high-flying corporate career without a spouse in a full-time supporting role. She believed the way she and her husband formed a team with him providing full-time support was the only way for a woman to rise to a high rank.

Many of the young women in that company sought to take a develop-mental assignment that involved a transfer at a convenient time for them. They didn't realize that what the company desired was a commitment to move anywhere at any time. Eve believed the younger women needed to under-stand the company's perspective, so she invited the Vice President of Human Resources to speak to a small group of young female technical professionals. After the meeting, word of it got back to Queen Bee, who was outraged. Queen Bee called Eve to berate her for daring to organize a meeting of women within the company without first seeking her blessing as the head of the women's network. She told Eve that she was never to speak to the Vice President of Human Resources again. From Queen Bee's perspective, the corporation's women's network was not a group run by women to discuss issues of interest to them, but rather a group over which she had complete control.

Eve shared her woes with a female colleague at the same level in the corporation as herself. Her friend described how that Queen Bee had attacked multiple women, who were far below her in the organization. That Queen Bee was so insecure, she saw every other competent woman as a threat and sought to neutralize them. Many of Queen Bee's male subordinates also hated her, but they benefitted by not being attacked as viciously as her female subordinates.

Queen bees in male-dominated organizations perceive other talented women as competitors to be eliminated. These Queen Bees are stuck in a mindset that there are limited positions for women and see other talented women as a greater threat than their male peers.

9.5 Queen Bees Are Everywhere

Maria Angela has dealt with Queen Bees all her professional life, and continues to do so, in her current consulting work, although now, observing from afar, and not affected by their stings. While she was working in the research center of PDVSA,[8] in Venezuela, Maria Angela became the prey of a woman, who was recognized as a Queen Bee, because she never gave opportunities to other women, only to her male friends in her department.

Luckily, Maria Angela was transferred to the headquarters of PDVSA in Caracas, where she was mentored by extraordinary female role models, some of them pioneers in their roles as presidents of joint ventures. Later in other countries, Maria Angela witnessed the way some female leaders limited opportunities for other female professionals and monopolized opportunities to represent the company or speak about diversity.

9.6 Rise and Fall of a Queen Bee

While consulting from Houston, Maria Angela had a case of a Queen Bee, which made a huge impression on her, because it replicated what she had repeatedly seen in many other companies, the rise and fall of a Queen Bee. It was in a joint venture, and Maria Angela had the privilege to mentor many

---

[8] https://en.wikipedia.org/wiki/PDVSA as of December 15, 2021.

employees, both male and women, who discussed during mentoring sessions the behavior of the Queen Bee.

This Queen Bee, "Dorothy", was one of the first women in the company. Not only was she a managerially capable engineer, but she was also very beautiful. She caught the attention of her team's supervisors, who fast tracked her career and promoted her into a supervisory role. Before her promotion, Dorothy had good relationships with her peers, and they thought her apparently caring attitude reflected her true feelings. After Dorothy became the company's first female manager, she became accustomed to being treated as an especially talented person and basked in the glory of being a pioneering woman. Unfortunately, it also prompted her to deny promotions and opportunities to other female engineers who previously had been her peers. She blocked the advancement of any potential female competitors, even those who were still at a very low level. No other woman had a chance to advance.

Soon, Dorothy was two-levels higher than any other woman in the company and reached an executive role in the C-suite. In her executive role, Dorothy was praised not only by members of her company but also by external stakeholders, who flattered her to get access to business opportunities. She developed an inflated sense of her own abilities, while denying opportunities to other talented women. Dorothy never shared opportunities to speak at international conferences and didn't sponsor other women for promotions, even if they had the technical and managerial skills required.

Maria Angela realized Dorothy was a manifestation of the Queen Bee syndrome. Deep inside, Dorothy did not want other women in leadership roles, because she felt it would devalue her standing as a pioneer. Unfortunately for the other female employees, the local culture in that country involved praising the leaders. Other women were trapped because Dorothy was blocking their progress to benefit herself. Eventually, Dorothy's superiors noticed the performance of her group was not as good as they expected. Dorothy loved being the Queen Bee, but the hive was tired because she had not nurtured younger talented women and men, so top management asked her to retire early.

Top male executives of the company told Maria Angela that "Dorothy was not a team player." Maria Angela suspects that the company's top leaders learned about the low morale Dorothy created around her, because instead of using her pioneering role to inspire and nurture other talent, she blocked others to maintain her own unique and prestigious role as the Queen Bee. Dorothy did not realize that eventually, people notice jealousy and failure to develop the talents of others. The new top leaders, even though they were all male and revered and respected Dorothy, decided her time was over.

Dorothy's behavior had long-term consequences for the company. Even after her retirement, younger women did not feel empowered to lead because they had not acquired the experiences that they needed to be leaders. The younger women did not have experience delivering presentations at major conferences or speaking as a member of a panel session. Maria Angela dealt with highly frustrated female employees who, even a year after Dorothy's retirement, cried when they told Maria Angela about how much they had done for the company, only to receive no recognition.

9.7 Queen Bees Attack Vulnerable Young Women

"Gabriella" came to the United States from Central America as a child and secured asylum status. After putting herself through college, Gabriella earned a degree in engineering and went to work for an oil company in a field office, where there was only one other woman, who was about twice Gabriella's age. Gabriella described how the other woman began by being nice but started "*turning on her and began putting her down.*" She said, "*She went from sweet to nasty within a month – Everything was a 'put down' – my accent, my hair – and after all I did to get there, she doesn't give me any important work to do. I was a zero.*"

Another woman we interviewed, "Belle" in United Kingdom was still emotionally fragile, because of her experiences with a Queen Bee almost two decades ago. Belle said, "*I no longer believe in rainbows, especially coming from women. I have become a very guarded individual in terms of trust and in terms of looking up to people.*"

Belle encountered the Queen Bee when at the age of 26, she accepted a high-level position in a very hierarchical governmental organization and was supposed to report to both the minister and the Queen Bee. She described herself as "*a very hard working and diligent individual.*" Even though the Queen Bee had far more experience and was twenty years older than Belle, she felt threatened by her and sought to block her. "*It came to the point, I couldn't take it anymore, burst into tears and went to see the minister and offered to resign. The minister, who was in his early 60's, said, 'Why do you want to do that child?'*".

After Belle explained the problems that she had with the Queen Bee. The minister responded, "*You know what child, those who have no enemies have no value. The more enemies one has, the more valuable one is.*"

Belle said, "*It took me several years to understand what he meant. I was doing a very complex job for that government ministry and just trying to do my best. To get all this 'shit' for no reason whatever. Back then, I didn't know the game – didn't know politics and a lot of things I now understand. I went into this sea of sharks with a very clean soul.*

Later, a colleague told me that the rumor was that I was the minister's mistress, but the minister was known to be a very proper man and the rumor was that I was his niece.

That experience was my first encounter with the political game that was difficult for me to explain to myself. I refuse to play it. I have made it my mission ever since not to do to others what was done to me and to make sure that I help as many people as possible so that they never have to encounter what I have…

My advice to young people depends on what you want to do with your life. If you are a believer in what doesn't kill you makes you stronger – put up with it and learn from it – we learn from bad experiences. It shows you the dark side of humankind and what people are willing to do and what you should be striving not to do. If you can, don't put up with it. If you are only looking for a paycheck or an easy life and not Maslow's theory,[9,10] *and self-actualization.*" (In Abraham Maslow's hierarchy of needs is that there are certain basic needs which must be met before other needs can be satisfied with the ultimate level being self-actualization.)

Almost two decades later when we spoke with Belle, she had still not fully recovered from her stressful experience with the Queen Bee. She cried while she was explaining what happened to her. No one should ever suffer from bullying. When you recognize that an abusive supervisor is adversely impacting your mental health, get out of the toxic position as quickly as you can.

9.8   How to Recognize a Queen Bee

To prevent a Queen Bee from abusing her staff, upper management must recognize the signs early and act. Below is a list of warning signs that an executive is manifesting the Queen Bee Syndrome.

- Offers fake praise to employees, which does not include rewards

---

[9] https://en.wikipedia.org/wiki/Maslow's_hierarchy_of_needs.

[10] https://www.eajournals.org/wp-content/uploads/Abraham-Maslow%E2%80%99s-Hierarchy-of-Needs-and-Assessment-of-Needs-in-Community-Development.pdf.

- People feel her behavior is insincere and without substance, and that she monopolizes all attractive opportunities
- Female employees are afraid to approach her

and she does not:

- Mentor other women in the company
- Offer development opportunities to female employees
- Help identify future female leaders
- Sponsor other women for advancement to higher roles
- Share external opportunities to serve on committees, panels, and other roles to publicly represent the company
- Groom junior employees to advance
- Empower her female employees

9.9 Imposter Syndrome

Queen bees appear to be a manifestation of the "Imposter Syndrome." People suffering from this mindset have risen to a position of authority in their organization, but still feel insecure. In 1978, when women were gaining footholds in the workplace, Pauline Rose Clance and Suzanne Imes[11] described how a person suffering from the "Imposter Syndrome" fears that they only succeeded due to luck, and not because of their talent or qualifications. The impostor phenomenon occurs among high achievers who cannot internalize and accept their success. They attribute their accomplishments to luck rather than to ability.

Clance and Imes thought only women suffered from Imposter Syndrome, but subsequent investigators determined the syndrome is not gender specific. According to the American Psychological Association,[12] anyone who differs from most of their peers—whether by race, gender, sexual orientation or some other characteristics—is more likely to suffer from the syndrome and feel like a fraud. People who are shifting into a new discipline or role are also vulnerable to feeling like an imposter.

The Imposter Syndrome undermines efforts to increase diversity in the workplace, because the first people from underrepresented groups chosen for advancement may perceive other people from their minority as their primary competition. To protect themselves, these token minorities may

---

[11] https://psycnet.apa.org/record/1979-26502-001.

[12] https://www.apa.org/gradpsych/2013/11/fraud.

unconsciously or consciously hold others who are like themselves to a higher standard of performance.

Whether or not the person with imposter syndrome is in a position of authority, working with them is much less pleasant than working with people who feel confident about their ability to contribute. Peers with imposture syndrome may be difficult because they are hesitant to make decisions and are reluctant to defend their conclusions.

9.10   Avoiding the Bee's Sting

If you sense you are making someone uncomfortable, back off and give them time and space to adjust. Don't try to demonstrate your superiority at someone else's expense. Give the difficult person time and space to adjust and get comfortable with you. Be secure in your own worth and don't push others into their discomfort zone. Avoid threatening the Queen Bee. If you have attracted a Queen Bee's attention, give her a way out.

When Eve was being attacked on the telephone by the Queen Bee for organizing the meeting with the Vice President of Human Resources, she truthful told her that she had just been informed her mother had died and was rushing to wrap up a few things before leaving the office. The Queen Bee instantly morphed into a sweet Southern Belle and began spouting condolences.

If possible, distance yourself from insecure leaders. Working with and for people who feel secure in their own abilities is much easier and more enjoyable. Find confident, competent people whose skills complement yours. That kind of team really can be self-managed!

10

Unconscious Bias

We can both suffer and benefit from unconscious bias. Bias[1] is prejudice in favor of or against a person or group and is often considered to be unfair. People who benefit from unconscious bias[2] may be unaware of the advantages they have and may attribute their success purely to their own skills.

Unconscious bias is often a significant factor in hiring, business opportunities, and promotion. Ideally, we would all be unbiased, but various forms of bias are constantly present. Perhaps the most famous case comes from symphony orchestra audition results analyzed by Claudia Goldin and Cecilia Rouse.[3] If the judges had clues to the gender of the musician, men were more likely to be selected. The percentage of women in orchestras did not significantly increase until those making the selections could only hear the applicants' performance. Any clue that the applicant was female, including the sound of high heels walking across the stage, skewed the judges' evaluations to be more negative.

All of us are biased in some ways based on our upbringing and experiences. We may be aware of some of our prejudices, but not of all of them. Even the most impassioned feminists may discover that they have an unconscious bias about women in leadership roles. Maria Angela was shocked to find this

---

[1] https://www.merriam-webster.com/dictionary/bias.

[2] https://diversity.ucsf.edu/resources/unconscious-bias.

[3] Goldin, Claudia and Cecilia Rouse, January 1997, US National Bureau of Economic Research Working Paper 5903, Orchestrating Impartiality: The Impact of "Blind" Auditions on Female Musicians, https://doi.org/10.3386/w5903. https://www.nber.org/papers/w5903.

© The Author(s), under exclusive license to Springer Nature Switzerland AG 2022
E. Sprunt and M. A. Capello, *A Guide to Career Resilience*,
https://doi.org/10.1007/978-3-031-05588-1_10

was true about herself. She was with her family when one of her nephews had surgery. When the neurosurgeon entered the room, Maria Angela realized she had incorrectly assumed the surgeon was a man. Seeing the female neurosurgeon prompted Maria Angela to reflect on her own deeply buried unconscious biases.

10.1 Spousal Consent

One outdated form of unconscious bias that adversely impacted both Maria Angela and Eve was their employers seeking their spouses' consent about situations that could impact both the employee and her spouse.

Early in her career, Eve was chosen to make a presentation when her company's Vice President of Research from the corporate headquarters in New York, was visiting the Dallas research facility, where Eve worked. The evening following her presentation, Eve and her spouse were invited to dinner at one of the fanciest restaurants in Dallas. At the beginning of dinner, the Vice President turned to Eve's husband and asked, "Does Eve like New York?" he answered truthfully, and said, "No." Eve's husband didn't realize that the Vice President was asking if Eve would be interested in working at the corporate headquarters. A move to New York would have been very disruptive to Eve's family, but it should have been her decision, not something that was sprung unexpectedly on her husband during dinner.

Eve would have preferred to have an opportunity to evaluate the pros and cons of taking a position at corporate headquarters in New York. Later, she learned to avoid bringing her husband to similar events where he might end up being the decision maker instead of her. She became very selective as to when she asked her husband to accompany her to any kind of business event.

During Maria Angela's early and mid-career years in Venezuela, her employer also sought spousal consent. In her case, she experienced this discriminatory practice about an academic scholarship. In 1975 the Venezuelan government launched by executive decree, the Foundation Gran Mariscal de Ayacucho, with the objective of encouraging Venezuelans to pursue advanced degrees, especially in STEM[4] (Sciences, Technological, Engineering and Mathematics). Long before Chavez came to power, the Venezuelan government was worried about a brain drain, so they offered scholarships or grants to promising young citizens so they could pursue

---

[4] Note—Some groups have added Medicine to the STEM acronym and use STEMM instead.

advanced degrees within the country or internationally. Tens of thousands of young Venezuelans earned their masters and/or doctorates through this program. PDVSA, the Venezuelan National Oil Company, adopted a similar program offering scholarships with the caveat that the graduates were required to return to work in one of the companies in the oil, gas or steel industries in Venezuela.

While Maria Angela was working in Intevep,[5] she noticed the employees around her age were treated differently depending on their gender: Men were offered an opportunity to pursue a Ph.D., while women were only given a chance to earn a master's degree. She realized this was a systematic, but unwritten practice which accompanied an unwanted paternalistic approach to managing female employees. At the beginning of the scholarship process, a Director of Intevep called Maria Angela's husband to inform him that his wife had been approved for a scholarship and asked his permission to inform her. Maria Angela's husband, Herminio, told the Intevep Director, "*but of course!*" However, Maria Angela thought that she should have been informed first and allowed to handle the situation. Even though family-work balance was a major concern because they had two daughters, one and three years old, Herminio agreed, initiating one of the most beautiful chapters of their dual career experience. In August 1994, Maria Angela started her first semester at the Colorado School of Mines as a single parent with their two daughters. Herminio requested a one-year unpaid leave of absence and joined his family in January 1995.

During her first year of studies at the Colorado School of Mines (USA), Maria Angela was ranked number 1 out of 553 students. She studied rock physics and 4D seismic becoming to her knowledge, the only student who started in August 1994 to complete a Master's in Science (Geophysics), in only one and a half years. When she received her degree in the graduation ceremony in December 1995, her mentor, Prof. Michael L. Batzle, and her tutor, Dr. Thomas L. Davis, sent formal letters to her employer, highlighting her outstanding results, and urging them to provide financial support so that she could pursue a Ph.D. degree. Without Maria Angela's consent, Herminio was asked by her company whether he thought extending her studies was a good idea and he enthusiastically said, "*Yes!*" However, despite her advisors' and her husband's support, the request to continue for a Ph.D. was declined, because Maria Angela was expected back at the company to lead several projects.

---

[5] http://goldmercuryaward.org/laureates/instituto-tecnologico-venezeolano-del-petroleo-intevep/, as of December 12, 2021.

When Maria Angela returned to Venezuela, and resumed working on January 1996, there were no projects for her to manage in the Earth Sciences Department, because she returned six months earlier than anticipated. Her supervisor, the Director of her Division and Human Resources, recommended that she take vacation time, and perhaps even an extended leave of absence. Maria Angela could not believe the situation. At least four of her male colleagues continued in their universities pursuing Ph.D. degrees in the USA and in the UK at the normal speed of four to five years. High performance, swift results, achieving a top ranking in her master's degree program did not earn Maria Angela the opportunity to pursue the Ph.D. she wanted. But her MS gave her technical and experiential self-confidence that opened a new phase in her career. She sought further certifications and reinvented herself.

At that time in Venezuela, there was a culture of placing a higher priority on family values than on business objectives. Family was perceived as solely the responsibility of women. Maria Angela's supervisors questioned her husband and got his opinion, but then chose to make their own decision based on their sense of where a woman should place her priorities. In retrospect, silence was not the best answer. Maria Angela should have challenged the decisions made by Human Resources and based her appeal on how her male colleagues were treated in similar situations.

10.2 Paternalism

The advice mentors and sponsors provide may be influenced by what they would like for their own children. This is termed "paternalism" which is a biased expression suggesting the advisor is a man.

Britannica.com defines paternalism as, "*attitude and practice that are commonly, though not exclusively, understood as an infringement on the personal freedom and autonomy of a person (or class of persons) with a beneficent or protective intent.*"[6]

A kindly, old-fashioned, benevolent and protective mentor may provide the same advice to female mentees that he would give to his daughters or granddaughters, but that advice may be career-limiting if he has a paternalistic attitude. Many career building experiences involve risk exposure. Such experiences may include travel to areas where personal safety is at risk and/or accommodations and hygiene are substandard.

---

[6] https://www.britannica.com/topic/paternalism.

Some paternalistic mentors may also discourage field and operational work, where hygiene facilities for women may be non-existent or primitive. The workers in those challenging areas sometimes pride themselves on their dangerous and/or difficult working conditions and may also be somewhat hostile to the presence of women, who they view as intruding and complicating things. Those men may be opposed to making any accommodations for women. Instead of seeing themselves as biased, they may perceive modifications made to encourage women to work in these areas in which they are under-represented as reverse discrimination.

Travel, even for a limited period, to some locations may expose a person to sexual, physical and health risks. While in the short term the person might feel avoiding some areas and tasks to be excellent advice, because it shields them from difficult and dangerous situations, in the long run it can be career limiting. If possible, companies and organizations should improve working conditions for all employees and mitigate risks. If a situation is dangerous or unhygienic for women, often it is for men too. No one in an organization should be unnecessarily exposed to risks.

If a mentor is concerned about how a mentee will handle difficult conditions, it is better if the mentor accurately and as completely as possible describes the sanitary, physical, and emotional challenges involved. That allows the mentee to decide and to prepare and plan for the difficulties. You, not your mentor or sponsor or your spouse, should be the one to decide which risks and hardships you will accept.

10.3 Managing Risks

In the early 1990s, Eve went to Nigeria on business. Even in good times, many people consider Nigeria to be a challenging place to visit. Eve was supposed to give a lecture in Lagos to company employees and then proceed to Port Harcourt and from there to an offshore rig for an experimental coring operation. Eve was responsible for the laboratory testing of a specialized coring fluid, which was being used for the first time on that offshore well. She was an expert in laboratory testing, not field operations, and had never been on an offshore rig. The lead production engineer, Mike, knew Eve had no offshore experience, but decided she should be on the rig to manage any real-time changes needed in the coring fluid.

Eve suspected that her presence on the offshore drilling rig was more for her education than for her value in providing real-time, on-site advice. A woman on the offshore rig would be a nuisance because there were limited sleeping quarters. Almost all the sleeping accommodations were shared, and company policy was that a woman should not sleep in the same room with men. No other women were expected to be on the rig, so Eve would either bump one of the top rig hands from his private room or occupy the space for two or more people.

To get to Nigeria, Eve flew with Mike from Dallas to London, where they planned to connect to a flight to Lagos. When they arrived in London, they learned their flight to Lagos was delayed by at least half a day, because of a general strike in Nigeria. During the delay, Eve and Mike met a couple of men, who were working for a service company and would be contractors on the rig where they were going to be involved with the coring operations. When the service company learned about the general strike, they told their employees not to board the flight to Lagos. Eve and Mike were not told to cancel their trip, so the two contractors also chose to wait for the flight. As service company representatives, the men thought it inappropriate to turn around if employees of the operating company were proceeding ahead.

When Eve arrived in Lagos, the general strike was still in effect. The electrical power was out, so, instead of lights, the runway was lit with flares. Air transportation between Lagos and Port Harcourt was curtailed, so Mike suggested they travel to Port Harcourt by car. The local drilling manager looked at Eve, who was in her early forties and said, "Not with her." Recognizing the risks and the benefits, Eve didn't dispute the drilling manager's decision.

The next day, there was a coup, and the company asked everyone who was visiting and not essential to leave the country. Eve opted to stay until the following week to deliver her lecture on the new coring fluid. That meant instead of leaving Nigeria with a group of company employees, she would exit alone. She asked the drilling manager what she should do if there was another coup during the four hours she expected to be at the airport between when the company guards left her at immigration and when her flight departed. The drilling manager's advice was, "Make friends fast."

Mike acted as Eve's mentor, in an unbiased and non-paternalistic way, when he pushed her to accompany him to Nigeria. That allowed Eve to choose which risks she was willing to take. The trip involved multiple risk assessments. First was whether to go to Nigeria. Having arrived in the middle of a state of emergency, Eve then had to decide what she would and wouldn't do and when to leave the country. The choices she made limited her physical risks and built her technical reputation, because she stayed to complete what she believed to be her most meaningful task. Fortunately for Eve, her trip home went smoothly.

Eve's time in Nigeria was not without frightening moments. She was the sole passenger in the van going to the airport with two gun-toting guards. Eve was terrified, when for some unknown reason, the driver exited the highway in a place in which there was not a paved exit. She wondered if she was being kidnapped but didn't say anything. She was tremendously relieved when her guards escorted her into the airport terminal and helped her check-in. She knew she wasn't supposed to pay any bribes, but when her guards asked her how much it was worth to her not to have her checked baggage opened, she asked the price and paid it.

At that time in that company, being willing to travel to Nigeria was an important way of demonstrating you were dedicated to your job. Even though she didn't go to Port Harcourt and out to the rig, Eve got credit for being willing to go and being present during a particularly risky time.

When Maria Angela was starting her career, there was a course on carbonates all new hires in geosciences wanted to take, because it included travelling to remote and beautiful locations on the Caribbean shore that were only accessible by private flights in small airplanes and boats. Maria Angela was thrilled to be selected to attend this course, but she had just learned that she was pregnant with her first child.

Maria Angela asked her supervisor what she should do. He explained this would probably be her only opportunity to take this course, because it was very expensive, and the company was considering eliminating it from the career training track. Maria Angela asked him for the logistical details, and he provided details about the time on boats and airplanes, adding that in his opinion, the risk was minimal.

Maria Angela also consulted with her obstetrician, who said because it was her first pregnancy, he could not predict how it would go and discouraged her from attending. Taking both her doctor and her supervisor's advice into consideration, Maria Angela decided not to go.

Her pregnancy went well, and Maria Angela does not regret her decision, even though as her supervisor predicted, the company cancelled the course, and she never had another opportunity to take it. She was able to make her own informed decision, balancing the risks involved with the benefits. The lesson learned here is when you are making an important decision, consult experts outside your organization to gather information on the risks to minimize future regrets.

10.4 Leverage All Information

In the mid-1980s one of Eve's colleagues, Ben, told her she was less valuable to the company than he was, because as a woman, she couldn't go to Saudi Arabia. Eve immediately challenged Ben's assessment by proposing a technical service project to Saudi Aramco. When it came time for the annual project reviews in Saudi Arabia, Eve was invited, and her company made the extra effort required to get a woman a business visa. She flew to Saudi Arabia along with Ben and presenters from other companies. On arrival, the immigration officer took Eve's passport, supposedly to check out a visa issue, and left her sitting outside an office. Ben and the other visiting expats spent several hours trying to get her admitted before the immigration officer returned her passport and let her into Saudi Arabia.

On her second trip to Saudi Arabia, Eve traveled alone and swiftly passed through immigration. After several trips with and without male colleagues, Eve realized if she traveled with male colleagues, the Saudis used her as a pawn to harass the men. If she went alone, the Saudi immigration officers processed her quickly and didn't raise questions about her visa.

At the end of her first visit, one of the visitors from another company warned Eve that on her departure while passing through security, she would be taken away and searched in private by women. He explained that the men at the inspection point were not allowed to check women. It was nice to be forewarned that she would be separated from the rest of the group. Having been warned, Eve didn't get worried when she was sent off by herself to a secluded room, where Saudi women were sitting around drinking tea and eating. As directed, Eve sat down, and waited until a woman came over and without speaking a word, patted her down. After that, the woman signaled that Eve was free to go to the departure lounge.

Information shared by both those who resent your presence in the workplace and those who support it can be equally valuable. Figure out how to make lemonade out of sour words.

On that first trip to Saudi Arabia, Eve's departure involved more irregularities. It was back in the days of paper tickets. When Eve checked in at the airline counter, the agent pulled her first-class ticket, but didn't give her a boarding pass. When none of the men she was with took her situation seriously, Eve phoned her local sponsor, who rushed to the airport, contacted a person in authority and got her a boarding pass. With her boarding pass in hand, Eve returned to the lounge, where the guys asked what her seat assignment was. One man realized he had the same seat assignment. Eve consoled herself that because it was a British Airways flight, if she got on board before that man, she would benefit from bias towards giving the seat to a lady. The man with the duplicate seat assignment was worried and rushed off to get another seat assignment. Eve never learned whether he was being upgraded or also had purchased a first-class ticket.

10.5 Enjoy the Benefits

If you are a member of an under-represented minority in your work environment, you face many barriers that those in the majority do not. When a sponsor helps you because of unconscious bias or a desire to help someone in your minority category, seize the opportunity.

Sometimes a person in a position of power notices something about you that prompts them to help you. Don't ignore these overtures. Welcome them! As a member of an under-represented minority, there is a lot greater scrutiny of everything you do, good and bad. Use the opportunity to showcase your abilities. Much of life is luck. When you get lucky, it is beneficial to figure out why, but while you are trying to determine why you have been chosen, grab the prize and don't forget to say. "*Thank you!*".

People in senior positions often are asked to recommend someone for something beyond their normal assignment. Depending on how busy the senior person is, they may provide the name of the first person who pops into their head, who comes close to meeting the criteria for the opportunity. If you are that person, be sure to thank the person who suggested you.

11

Cross-Cultural Issues

Our cultural heritage, upbringing, and youthful experiences, coupled with how others perceive or react to our style, shape who we are and how we see the world. We use our prior experiences to interpret our present reality. Differences between our cultures impact our ability to communicate effectively. Mentors do not have to be and often are not from the same culture or social background as their mentees, but it is important that they are aware and sensitive to how cultural differences impact the mentoring process.

Both mentor and mentee should enter the conversation with an open mind. Ideally, because mentors are in the power position in the relationship, they ought to strive to overcome differences in race, ethnicity, gender, age, and nationality to communicate effectively and provide unbiased advice. However, because the mentee is "imposing" on the mentor's time to get advice, the mentee should make the conversation as easy as possible for the mentor by paying close attention and being slow to take offense.

Both participants should strive to maintain a non-judgmental attitude. Active listening can enable us to detect underlying intent when the words mean different things to the mentor and mentee. Engage in these discussions with curiosity, an open mind, and a sense of humor.

If either mentor or mentee is confused, they should feel comfortable asking clarifying questions. It is better to cover fewer questions and to have an in-depth discussion of the key issues, including critical nuances, than to squeeze in numerous questions and risk miscommunication.

© The Author(s), under exclusive license to Springer Nature Switzerland AG 2022
E. Sprunt and M. A. Capello, *A Guide to Career Resilience*,
https://doi.org/10.1007/978-3-031-05588-1_11

One of the important factors in cross-cultural mentoring is the perception of relevance of the advice. The mentee should be prepared to discuss their priorities and constraints, so that the mentor can provide better informed guidance and more targeted alternatives. Both participants should pause to allow for clarification questions.

The mentor and mentee may have different perceptions of timelines and the priority that should be placed on meeting those timelines. What is urgent for one person is not for another. What needs careful planning and anticipation in the opinion of one person because of her or his culture or personality, can be handled with an impromptu and improvised style by another person. This can be a source of friction in mentoring sessions.

To have a respectful and smooth mentor–mentee relationship, be aware of what respecting time means for your counterpart and comply with their expectations. As a mentee, you should always arrive no later than the scheduled time for a mentoring meeting and expect to wrap up by or before the end of allocated time. Remember to thank your mentor.

11.1   Unexpected Issues with Same-Culture or Same-Race Mentoring

As human beings, we are all different. Even within a narrow cultural group, individuals may have different priorities, preferences and styles. Siblings in a family, even if they are born within a few years of each other, can have very different priorities and preferences. Eve was one of five children born within an eight-year time span. She and her siblings were all raised in the same house in Brooklyn, New York, but chose very different career paths. Eve has a Ph.D. in geophysics, but only one of her other siblings has a college degree. Her older brother dropped out of college after a few weeks and is a lifelong hippie and ski bum. Her younger brother has a college degree but is a highly skilled tool and die maker. One younger sister dropped out of college, had two kids, built a highly successful commercial real estate business, retired very young and joined the Peace Corps. The youngest sister was a subway train driver and later a realtor. We had very different priorities and chose divergent paths for our lives.

When the mentoring relationship is between people of the same culture, race or nationality, often the assumption is that the mentor and mentee share the similar values or background. Also, when mentees contact mentors, there is some preselection presumably because of the mentee seeing something in the mentor's background that they feel would make them a good source of advice. However, just as we develop different friendships, some mentoring relationships work much better than others.

Maria Angela observed that same-ethnicity/same-culture mentoring in the Middle East may not work as well as cross-ethnicity and cross-cultural pairings. When Maria Angela was in Kuwait and seeking interaction with people like her, she realized that being the same nationality did not necessarily mean sharing the same values. There was a huge divide between pro-Chavez/Maduro supporters and those who opposed Chavez and were part of the Venezuelan diaspora. When both sides were present at social events, the atmosphere was tense.

As the daughter of Italians in Venezuela, Maria Angela had a different Italian culture which separated from the culture in Italy when her grandparents moved to Venezuela. Maria Angela visited Italy on vacations but was surprised that the Italian women expats from modern day Italy in Kuwait did not know how to cook and could not speak English as fluently, as she did. Most of the Italian expatriates in Kuwait belonged to a new generation of Italians. As a multicultural "product" Maria Angela interacted with the fresh from Italy community and the Italian-Venezuelan one but found both groups very different from what she experienced in Venezuela.

In many ways, we are all unique. The more differences there are, the more challenging it can be to have open and honest communications in a mentoring session. We all must recognize that, in a mentoring session, the mentee should clearly articulate the questions on which they are seeking guidance. Mentors should pause frequently to check if they are providing the type of advice the mentee desires and whether the mentee has any questions.

Mentees should prioritize their questions, so that, within the limited time available, the most important issues are addressed. It is important to recognize that sometimes, the mentor they targeted is not the best source of advice for them. The mentee-mentor relationship is not a marriage. If for any reason you find the interaction to be unsatisfactory, do not hesitate to thank that mentor for their time and seek advice from someone else.

11.2 Adapt Your Style

During her fifteen years in the Middle East, Maria Angela learned to go with the flow and pay close attention to cultural nuances. She was asked to mentor one of the high-flyers of a company in the energy sector of Kuwait. This very young woman was an outstanding performer who, because of her ability, had been promoted faster than her peers.

After one of the mentoring sessions, Maria Angela's mentee invited her to accompany her to inspect the progress of the new building in which her team would occupy half of a floor. When Maria Angela accompanied this gentle but energetic woman to visit the building, she was shocked by her driving. Her mentee drove at a very high speed and entered the parking lot by driving in an exit the wrong way.

Once inside the building, which was a wonderful, modern office building, the mentee expressed astonishment. "*Do you see this, Ms. Maria Angela! This is unbelievable! We need to do something about it!*"

Maria Angela did not understand the problem. Her mentee was pointing at the restrooms area where there were signs for men's restrooms and women's restrooms. Maria Angela quickly reacted. "*WOW! It is something to do with the restrooms' entrances? Right?*"

Her mentee responded, "*We cannot have the women's restroom door close to the men's.*"

Maria Angela realized the close positioning in the building was culturally unacceptable in Kuwait. The solution was to have both restrooms at one end of the floor for women and the other two restrooms at the other end of the floor for men. This experience reminded Maria Angela about how important cultural nuances are in mentoring sessions and that mentors must pause and ask clarifying questions.

Another time, Maria Angela mentored a very capable and smart Kuwaiti national on what to do during his interview for the Harvard Business School (HBS) leadership program. Like Maria Angela, he was an alumnus of the Colorado School of Mines. Kuwait Oil Company (KOC) had a program to send valuable people like him to the HBS Leadership program, but KOC's candidates had to pass the HBS admission process, which was based on their resume and an interview. The interview was important.

When Maria Angela was coaching him, she noticed he was not looking directly into her eyes. She explained to him that she understood that looking directly into the eyes of another person was not culturally acceptable in Kuwait, especially if a man was looking at a woman who was not part of his family, but during his interview, he had to try to do it.

He explained to Maria Angela that in Kuwait, kids are taught not to look into the eyes of anyone, because that is confrontational. He said that at school, looking directly at another kid during recess, was a call for fighting and they went at it with fists and kicks! If he looked at his father directly, his father slapped him gently.

Maria Angela had a serious challenge training her mentee to look directly at people during the interview, because in the United States avoiding the eyes of the person you are talking to signifies that you are hiding something or that you are excessively shy or uninterested. In the United States, we describe someone who avoids looking at your face as "having shifty eyes." After a few rehearsals with her, the mentee succeeded in maintaining eye contact and won entry to the HBS program.

Another crucial point about cultural nuances is personal space, which varies significantly between different cultures. In the Middle East, when people are with others of their own gender, they cluster very close together: Women sit, walk, and engage in all activities among themselves in close proximity, touching arm to arm, even if they barely know each other. This was foreign to Maria Angela during her early years in Kuwait, when her new Kuwaiti friends would, for example, walk in the mall with her, holding her by hand to express friendship.

She felt it was wonderful to be able to express affection so openly. That was not completely foreign to her, because Italians are also very open in expressing their affection. However, in Venezuela, that is not well regarded, and in the United States, members of the same sex touching each other may suggest more than friendship.

For Maria Angela, adapting to what she perceived as the "huge personal space" that she noticed Americans require has been surprisingly difficult. She finds herself immersed in "*Excuse me,*" and "*Pardon me*" requests when picking items from the shelves of the supermarket, or when in line to pay at the pharmacy. Since she relocated to the United States during the pandemic, she may also be experiencing the greater distance people want to avoid contagion. Eve is acutely aware that acceptable interpersonal distances have greatly increased between people. You don't want anyone "*breathing down your neck.*"

Similarly, touching another person for greetings or during conversations differs between cultures. Europeans, like Italians, French or Spanish people, greet each other with hugs and two kisses. Venezuelans and Latin-Americans in general, also hug each other but only kiss once. Americans do not touch much, but handshaking is common among women and men. Even that has diminished greatly during the pandemic, with many just waving at each other across a six-foot distance. Middle Eastern men will never touch women when greeting them, and vice versa. Asians do not touch at all unless they know each other very well. Maria Angela adjusts her greeting style depending on who is in the group and experiences the joys of each culture as well as her own.

Now, with the Covid-19 pandemic, sensitivity about physical proximity has evolved. With the new health concerns, people may inquire before meeting about your vaccination status, current health and attitude about wearing a mask.

The pandemic has rapidly changed many things. Mentors are often older and may have health considerations that make them reluctant to be in close contact with people outside their household. Fortunately, we have many alternatives in how to "meet" for a mentoring session. Don't be offended if your mentor requests a video conference meeting or telephone discussion instead of a face-to-face meeting.

If you encounter someone at a public meeting, remember that not all people like to be touched, or even to stand close together while speaking. Maintain your distance at the beginning and avoid any physical contact until you learn what works well and what would be welcomed by the other person. As a rule, a mentor should never engage in any touching beyond initial greetings, and limit that touching to a handshake. It is better to be cautious and take a safe approach to the relationship.

Dress codes have a special relevance in mentoring relationships, and those are also very much culturally driven. Women have many alternatives in how they dress and face large cultural differences. Cultures have drastically different expectations about women's appearance and modesty. In the Middle East, women are careful to dress very modestly, usually with long sleeves and loose-fitting clothes. In Latin America, women have more freedom in their apparel, which may be openly sexy. Be conscious of the impact of how you dress, especially if you are a woman, because it can send misleading signals to members of a different culture other than your own. Eve had a senior professor at the University of Texas complain to her about how difficult it was for him to concentrate when young female Latin American students came to his office for advice wearing low cut blouses that enabled him to look down their cleavage.

Wearing classic, conservative clothing is always a safe way to overcome cultural differences. It always helps to mimic what you perceive as "normal" in your surroundings. It is not a matter of changing your essence but maintaining your style. Be respectful of cultural nuances, and always keep building and maintaining your precious reputation.

Unfortunately, dressing modestly doesn't always work. One of our confidential sources, "Gabriella," described how after she got married and gave birth, her breasts were huge because she was breast-feeding. Despite wearing modest clothes, the much older engineer, who was her mentor, began making offensive, dirty jokes and touching her inappropriately when they were alone driving to a work site. Men should control themselves, but women should also do everything they can to make it easier for men to do so.

11.3 Improving Cultural, Multicultural, Cross-Cultural and Intercultural Awareness

Cultural awareness and curiosity are beneficial because it helps us to overcome unconscious biases and preconceptions that we may have ingrained in ourselves because of having worked with similar nationalities or ethnicities in the past. Someone's culture is not always apparent by just the way they look. Maria Angela, for example, is a person who, even if she appears to be white and American, is Italian and Venezuelan, and has lived in the Middle East for 15 years. She classifies herself as a Latina and Italian. She has learned to adapt her style to the setting in which she is working, because what is acceptable in one culture is not in another. Even different organizations and regions within the same country may have significantly different cultures.

Mentoring sessions depend on conversations, so it is important to recognize that different cultures handle conversations in very different ways. For example, cultures differ about whether it is acceptable to interrupt someone during a conversation. In Italy and Venezuela, people interrupt each other frequently to show interest and indicate that they are following the conversation. If you wait patiently saying nothing, that often triggers a comment of "*What? You are not listening to me!*" In the USA, interrupting a person is terrible manners and listeners roll their eyes if the speaker continues talking, without paying attention to the person who has interrupted with a clarification or question. Often the person who interrupts is perceived to be trying to show his or her greater knowledge.

Another example related to enormous differences between Latin America and the Middle East, are smiles. In Latin America, people, even if they are strangers, smile at each other, but in the Middle East, people do not smile if they do not know the person. This was especially difficult for Maria Angela in close spaces, such as elevators. Growing up in Venezuela, where people would greet everyone on entering a bus, the metro, or a classroom, it was difficult to be inside an elevator and not say a word or see a smile from anyone. With time, Maria Angela mastered the skill of when and when not to greet people in elevators and other places. The custom also differs in the USA, where she is now living.

With the pandemic, there is also mask culture, which varies from person to person and region to region. When you meet with a mentor or sponsor, be prepared to comply with their health concerns.

11.4 Tips to Enhance Our Multicultural Communications

Not everyone is immersed in a multicultural environment, but we can all learn to communicate better. If you live in a small community with little diversity of ethnicity or cultures, you can still enhance your multiculturalism. We have found that sharing these tips with mentors and mentees enhances appreciation of the impact of cultural differences in mentoring.

- Be open-minded about differences. Value them because they enrich our society and our own professional and personal growth.
- Be sensitive that there are many other cultures besides your own and that other cultures are as valid as yours.
- Engage in conscious efforts to expand your network to include people from different cultures.
- Be curious and learn about the geography, history, art, music, food, and economic issues around the world. Greater appreciation of other cultures enhances our ability to bridge cultural differences.
- Develop an appreciation for diversity and multiculturalism.
- Attend multicultural awareness training.
- Learn about different languages and cultures.
- Recognize body language can be different in different cultures.
- Be curious about the meanings of gestures, words and phrases varies in different cultures.
- Be flexible if something unexpected or strange occurs during a mentoring session. Do not over-react. Stay calm unless you are physically or sexually assaulted. Consult with others about the unexpected behavior and guidance on whether to continue a relationship with that mentor.
- If you are physically or sexually assaulted, seek help immediately. End the mentoring session and get away from the aggressor. Report the behavior to the appropriate authorities as soon as possible.

Part IV

Keys to Success

That which is not good for the bee-hive
cannot be good for the bees
—**Marcus Aurelius**

12

Networking

Your network is your safety net. No matter how well you are doing within your organization, be sure to build a network that extends far beyond its borders. Consider building and maintaining your network to be an essential part of your career plan.

If you joined "Facebook" years ago, you probably marveled with Eve and Maria Angela, that it enabled you to get in touch with people you once knew, but you were no longer were able to contact. Through social media, you were in touch again with people from your high school, if not elementary school years.

Facebook is no longer new, and you probably communicate with very few of those long-lost friends. Why? If you lost contact again, there must have been reasons. Networking is not about the past. It is about the future, and you should take a strategic approach to building and maintaining your network. Some of the motivations include:

1) To broaden your outlook on your career and your life
2) To exchange ideas
3) To enhance your image in your profession
4) To connect with experts
5) To gain access to potential sponsors

If you regularly post information of interest to your social media contacts, they are more likely to think of you, when an opportunity of interest comes along.

Think carefully and strategically about everything that you post. Once something is out on social media, it never goes away and can come back to haunt you. Don't ever post anything anywhere that you wouldn't want a potential future employer seeing.

Creating a strong network is like planning, planting and tending a garden. You want a variety of plants that do well under different conditions and bloom at different times. You also want some larger, dramatic plants that get attention. You should regularly fertilize and water your garden to keep it healthy. On-going maintenance is required to keep it at its best.

Prioritize getting to know people in your organization. You don't want a reputation as a person who hangs around the coffee machine or water cooler all day or wastes time during video conference meetings or phone calls, but you need to develop relationships with people. Take a few minutes to greet people when you arrive or begin a call. If there is a lunchroom or cafeteria, eat lunch there, not at your desk. You can learn a lot about how your organization really works, who people like and who they hate, who is a rising star and who is on the way out. Whenever possible, help your colleagues.

12.1 Networks Must Be Constantly Nurtured

How strong is your business network? When was the last time someone in your professional network called you to ask for a favor? When did you do something important for one of your professional contacts? Networks that are not actively used tend to wither and lose effectiveness.

Eve learned the hard way that networks can quickly lose their effectiveness. She was feeling on top of the world when she finished her term as 2006 President of the Society of Petroleum Engineers (SPE). Her company, Chevron, had generously supported her, so she could travel all around the world in 2006 speaking to members of the society. At the beginning of 2007, while she was still in the role of Past-President and on SPE's Board of Directors, she went in for her annual physical. She had not been ill the entire previous year, despite a grueling travel schedule and believed herself to be very healthy for her age. Then her doctor peeked in her right ear and asked, "*Does your ear hurt?*".

A red spot on her eardrum turned out to be the tip of a medical iceberg. After a year of multiple major surgeries, the expert in her rare metastatic cancer gave her a 45% chance of surviving for five years, with the starting point backdated to when the red spot was first observed in early 2007. Her key sponsor at her company had retired, the boss she liked had quit, and her supportive boss's boss had also retired. Eve had switched employers in 2000 and wasn't yet vested in her company's retirement program. Both her and her husband's healthcare came through her employer. Then her new boss tried to bully her into quitting. In one year, through no fault of her own, Eve went from expecting a bright future to feeling crushed.

Just when she needed a strong network within her company, her network was in terrible shape. She had been with the company for seven years, but much of the previous year Eve was in the hospital or recovering from surgery. The year before that she was traveling the world as the President of SPE. On top of everything else, Eve was emotionally and physically fragile from her battle with cancer and her dismal prognosis.

When her new boss tried to persuade her to quit during their first meeting, Eve felt vulnerable, but leveraged her remaining network. Her assistants were people she had rescued when others within her network told her they needed new positions, because their roles had been eliminated. Both provided valuable insights and support. One of them shared that Eve was not the only person her new boss threatened. The assistant had learned through her grapevine that the newly hired boss was trying to replace as many of his direct reports as possible with friends from outside the company.

Eve took the risk of discussing with her sponsor's replacement what happened in her first meeting with her new boss. He said he would investigate it and urged Eve to speak to the company's ombudsperson. Six months later, the bad boss vanished with no announcements and Eve moved on to a new assignment.

She realized that when you have a rare cancer, the statistics are poor and the predictions of even the most knowledgeable experts can be wrong. She soothed herself, by recognizing that the expert based his grim survival prediction on studies of young men. She was not male and in her mid-fifties, she was no longer young. It was a reminder however, to make the most of everyday and to not put off to the future anything she really wanted to accomplish or do.

To her delight, six years later when Eve asked the medical expert what his new prognosis was, he refused to give one. He limited himself to saying, "That doesn't apply anymore." It was the best news he could have given her.

12.2 Networking Remotely

Nowadays, with many people working remotely, the concept of "hanging out" with your colleagues at a physical work location has practically disappeared, because there is little or no opportunity to interact in person. So, log into your meetings early, before the starting time. Plan your meetings with 10 to 15 min separation, so that you may ask some participants to "stay" longer with you for more intimate or one-on-one conversations.

A trick that has worked well for Maria Angela when working remotely is to link to formal working groups or teams a "WhatsApp" or "Slack" group platform. These are informal platforms on mobile phones that enable more direct and spontaneous interactions. Also, she makes a point of meeting her teams and/or colleagues in person for coffee or lunch if they are in her area or for breakfast or coffee at one of the global conferences she is attending.

Maria Angela has developed networks to promote women's empowerment around the world. She meets with members of her network at the annual conferences of multiple professional associations including the Society of Petroleum Engineers (SPE), the American Association of Petroleum Geologists (AAPG), and the Society of Exploration Geophysicists (SEG), and at annual gatherings in different parts of the world of her alma mater, the Colorado School of Mines, and her high school in Venezuela.

When you work remotely, it is even more important to network. Fortunately, now with widespread video conferencing, people are not just a voice, but also a face. Make the effort to get to know your co-workers and what is important to them. Consider a few minutes of social chitchat to be part of developing important business relationships.

Call people up instead of just messaging them. That gives you an opportunity for some personal discussion to get to know each other better.

A network that includes people beyond your current employer is extremely beneficial when you are looking for a new job. Even if you are completely happy with your job and have fantastic mentors and sponsors at work, you should develop a network beyond your organization.

You never know what the future will bring. Some of the largest and seemingly most successful companies fall on hard times and shrink dramatically. General Electric (GE) is an example of a formerly highly admired, leading company that has shrunk dramatically.[1,2] When Jack Welch was Chairman and CEO of GE between 1981 and 2001, everyone thought he was brilliant. The company's market value soared from $12 billion in 1981 to $410 billion

---

[1] https://en.wikipedia.org/wiki/Jack_Welch as of December 14, 2021.

[2] https://en.wikipedia.org/wiki/General_Electric as of December 14, 2021.

when he retired twenty years later. As of May 31, 2021, GE's market value was $120 billion, which adjusting for inflation, is only about 20% of what it was when Welch retired.

Other huge, highly respected companies that have shrunk dramatically in value include AT&T, Sears, General Motors, and ExxonMobil. At one time, all these companies were highly regarded and considered to be great places to spend your entire career. Building a strong network beyond your organization is important because you can't predict everything the future will bring and when you will need to look for another position.

Professional societies are a superb way to get to know people across many organizations. To really benefit from professional societies, volunteer to help. This provides an opportunity to not just meet people working in different organizations, but to work with them. If someone has had a favorable experience working with you on professional society projects, they can serve as a reference and write persuasive letters of recommendation.

If there isn't a professional society of interest to you and you feel that there is a need for one, don't be afraid to create one. In the mid-1980's at the celebration of the Society of Professional Well Log Analysts' twenty-fifth anniversary,[3] the keynote speaker described the motivations for forming the professional society. Listening to that speaker, Eve realized all those reasons applied to the niche discipline in which she was working at that time, core analysis. The long-time friend, who was sitting next to Eve, was also working in core analysis, but employed by another company. She suggested they should create the Society of Core Analysts.[4] It still exists and Eve is proud of having identified a need and filled it even though she moved on to other types of work. She hasn't attended a meeting of the Society of Core Analysts for about thirty years but remains in contact with some of the people she met through that group.

Eve and Maria Angela became acquainted by working together on professional society projects for the Society of Exploration Geophysicists[5] and for the Society of Petroleum Engineers.[6] Even though they have never lived or worked in the same city at the same time, they have gotten to know each other well and have supported each other in many initiatives.

When you attend professional society meetings, the social functions are as important, if not more important, than the technical sessions. Even if you don't know anyone in the group, you should attend social functions and

---

[3] https://www.spwla.org/.

[4] https://www.scaweb.org/about-the-sca/.

[5] https://seg.org/Default.aspx?TabId=176&language=en-US.

[6] https://www.spe.org/en/.

introduce yourself to people. It is better to attend social functions alone as opposed to with people you work with on a regular basis. The tendency if you attend with colleagues is to socialize primarily with them. If you go alone, you are more likely to start a conversation with strangers, who may later be valuable connections.

Many local groups have luncheons with time to socialize before the presentations. To maximize the networking value of the meetings, go early and talk to people you don't know. The first few times are the toughest, but it is a great way to expand your network. Be sure to have business cards and hand them out when you introduce yourself.

During the pandemic, many groups have been including video conference time for people to socialize. With video conferencing, you can get some of the benefit of face-to-face mix-and-mingle sessions. During the pandemic, Eve became better acquainted with members of her local chapter of the California Writers Club[7] and was asked to help implement her suggestion to create a speakers' bureau for members. Subsequently she was urged to take on some leadership roles.

After living in the Middle East for 15 years, Maria Angela was accustomed before the pandemic to working remotely with an extensive group of people in around the world to promote her volunteer initiatives with the United Nations, SPE, SEG and AAPG.[8] She has participated in meetings at awkward times, because back in 2008 and 2009, for example, no one was particularly aware of taking into consideration different time zones when scheduling meetings, and most of her teammates were in the USA, Latin America or Western Europe. It was fun but also challenging for her, but her fuel was a willingness to collaborate, and she acquired many mentors and sponsors for her work many of whom were thousands of miles away. She values this experience very much. For her, remote working was already her usual way to work.

12.3   Super Networkers

Your ability to network is enhanced when your contacts include a few "super networkers". These outgoing individuals constantly add people to their network and seem to know everyone in their profession. Even if they are not in a powerful position, they can be extremely helpful in suggesting who you should contact and by introducing you to other members of their network.

---

[7] https://calwriters.org/ as of December 2021.

[8] https://www.aapg.org/about/aapg/overview.

Be sure to include at least one "super-networker" or "hyper-connector" in your own network. If you read their descriptors, you already probably know who they are.

Networks that are not used decay. A healthy network requires continual maintenance. Contact key people regularly. When they call you for help, do whatever you can even if it you are incredibly busy. Do favors never expecting one in return. Pay it forward and hope that you never need help yourself.

12.4 Meeting Potential Sponsors

Sponsors are influential people who can vouch for your ability to do well in various roles. The best sponsors are well-networked people who have seen you in action and can vouch from personal experience that you can perform well in a role of interest.

The first step to finding a sponsor is to get to know influential individuals. Social functions of professional groups are great ways to meet potential sponsors. Don't pass on the social functions because you are tired from traveling or want to catch up with your email. Also, resist the temptation to hang out with people who you see routinely. This is your opportunity to make new contacts. Even if you are tired from a long day of meetings and have another long day ahead of you, prioritize attending the social functions and meeting new people.

Eve made an important reconnection at a social function of a professional society she rarely attended. A tall guy came up to her and asked, "Remember me?" It was the ultimate awkward moment. Years before, he was a graduate student at MIT in the same department as Eve, when she was an undergraduate. When Eve was a student at MIT, the ratio of male to female students was about twenty to one, so the guys remembered the girls and not so much vice versa. That turned out to be a tremendous advantage. Eve's meeting with the tall guy was a fateful reconnection. Don was at that time, the Chief Technology Officer of his company, Chevron. Over the next few years, Eve interviewed Don several times for articles she was writing for the Society of Petroleum Engineers flagship journal. When Eve's employer, Mobil, announced the intent to merge with a bigger company, Exxon, she emailed him about advice for what she might do because "the status quo was not an option." Her short note led to an invitation to interview and subsequently an attractive job offer.

When looking for work, it is better to ask for advice than to ask for a job. Everyone can offer advice, but people do not always have a lead to a specific job. You keep the dialog going by asking for advice.

When your organization is in turmoil because of a merger, it is a great time to look at your alternatives. You don't need to explain why you are interested in a new job, because for reasons unrelated to your performance, your current organization is undergoing a major change. Eve recognized that she would probably have to relocate and decided that if she looked for a job beyond her current employer, she would have more control over what she would be doing and where she would be based.

Eve's email led to a dialog and a job interview. When Don offered Eve a job, he said that she had "*shown the ability to reinvent herself and he was giving her another chance to do it.*"

Maria Angela considers her network to be one of her most important assets. When she began working in Kuwait, calling people was a challenge, because no one knew her. She made building a network to be one of her top priorities, and that helped her during her years of work in the Middle East. She learned from this experience and worked hard to maintain her networks in Latin America and in the USA. She never again wanted to be in the situation she faced during her first months in Kuwait.

12.5   How Robust is Your Network?

Building a network is building your future. Think strategically when analyzing your network:

- Does my network include people in a wide range of technical disciplines?
- Does my network include people with different levels of experience and prestige?
- Is my network well populated with contacts in different geographical regions and companies?
- How many potential mentors do I have in my network?
- If I were to lose my job, do I have connections in my network who can help me find a new job?
- Do I have the connections in my network that I will need in my next role?
- Do I have someone in my network who I can ask difficult organizational questions?
- Do I have friends in my network, who can provide emotional support?

12.6   Resist Your Inner Introvert

Make a pledge to yourself to engage in networking every day. Take a break from your work to chat with colleagues. If you are working remotely, call someone just to chat or to ask for advice. Attend a video conference social meet-up or go to a professional society meeting. Start in some way and do it regularly. Even after you have built a good network, you must constantly nurture and renew it, or your network will decay.

Think of your networks as living creatures that require nourishment, grooming, and regular exercise. Be sincere and honest with your connections. People are very good at detecting fake interest or fake support even if they are very far away.

12.7   Your Emotional Support Network

Another key component to personal resilience is an emotional support network. We all need emotional support and honest advice. To get all the help we need, we usually need a team, because our problems are often multifaceted and no one person can provide all the insights and help we need.

When Eve felt threatened by her new boss right after her yearlong intense battle with cancer, she didn't feel that she had anyone within the company with whom she could comfortably share all her fears. After that diffi-cult period, Eve became good friends with some colleagues, who provided valuable insights into the corporation's culture that enabled Eve to defuse situ-ations before they became serious problems. When you feel under attack at work, getting an insider's perspective on the situation can put your personal situation in perspective. For example, Eve learned that a high-level female executive who lashed out at her had a reputation for attacking other women and was widely hated.

During her career with Mobil, Eve's colleague, Nizar, was a very important mentor and he and his wife, Paula, provided valuable emotional support. With his knowledge of the corporate culture, Nizar could give advice that was better adapted to the company than outsiders like her husband. Nizar and his wife helped Eve through several crises and remain close friends.

Family and friends outside your workplace can provide important emotional support, but they can't put your problems in context the way a work colleague can. Having trusted confidants at work enables you to take a more strategic approach to resolving your problems. Trusted co-workers' knowledge of both the aggressor and the organizational culture enables them to provide more strategic advice. These advisors can explain how what you are suffering is part of a larger pattern. They can help you defuse the situation before it becomes a crisis. Just "venting" to someone who knows the organization can allow you to calm down and move on.

When you have been verbally attacked, and you learn the aggressor has a reputation for treating other people the same way, it can be easier to tolerate and manage. Often it is comforting to know that it wasn't so much what you did, but that the other person has a reputation for being difficult.

As an Italian, Maria Angela experiences with high degrees of emotion everything that affects her work and life, and she has learned the hard way that revealing emotions to everyone is dangerous. People are quick to label others "emotional" and other terms. Be careful about sharing your emotions. Have someone to whom you can brag about your successes without excusing yourself, because you know this person will celebrate the victory as much as you do. Pick your supporters carefully. Maria Angela's family provides her first line of emotional support. She shares unfiltered conversations about her troubles with only a few friends.

13

Resilience

Merriam-Webster Dictionary defines resilience as[1]:

1. the capability of a strained body to recover its size and shape after deformation caused especially by compressive stress
2. an ability to recover from or adjust easily to misfortune or change

Many aspects of work and life can cause intense emotional stress. Emotional stress can lead to physical stress and real illness. To survive and thrive we need to be resilient and surmount the mental and/or physical problems we encounter. We must learn from our misfortunes and bounce back stronger and more able to take on the challenges that life throws our way.

Over the course of a multi-decade career, we need the resilience to reinvent ourselves multiple times, because of rapid changes in technology and environmental concerns. When Eve finished her Ph.D. at Stanford in the late 1970s, she joined the oil industry. At that time, people from all the top universities saw a career in the oil industry as an avenue to help solve the energy crisis. Now, the petroleum industry is trying to reinvent itself.

Maria Angela started her shift towards sustainability in 2016, when she became aware of the United Nations (UN) Sustainable Development Goals (SDGs) through a contact with the Kuwait UN International Organization for Migration Office (OMI). When she saw the colorful icons representing

[1] https://www.merriam-webster.com/dictionary/resilience..

each of the seventeen goals hanging in the atrium of the UN building in Kuwait, she decided to learn more about sustainability. She educated herself about sustainability with a focus on a global scope and network and led the creation of the Geophysical Sustainability Atlas,[2] that maps geophysical techniques and methods to each of the seventeen UN sustainable development goals. This atlas created opportunities for Maria Angela, who also engaged in two certification processes, one with the Institut Francais du Petrole,[3] (IFP) and the other with the University of Cambridge in the UK. This prepared her to shift to an entrepreneurial consulting phase. She is pleased with her new career path that was initiated by securing residency by merit in the USA, and moving to Houston in late 2020, as soon as Kuwait eliminated the pandemic travel restrictions imposed on expats. Many people in the oil industry are attempting to find refuge in environmental work, but transitions are not always easy. The petroleum industry was able to attract top talent by offering attractive compensation, but employment in environmental work has typically not been as lucrative.

In contrast with other dramatic shifts in her life and career that were triggered in reaction to crises, this time, Maria Angela proactively planned her move leveraging her rules on how to become resilient:

- We are not born resilient. We become resilient.
- Focusing on long-term goals makes us more resilient, because short-term issues are minimized in comparison with our long-term objectives.
- Incorporate stress reduction routines in daily life, such as exercise, music, hiking and other activities to reduce stress, improve your physical well-being and prepare yourself to face adversity.
- Understand your strengths and weaknesses, so you can leverage your strengths, and resiliently overcome adversity.
- Build a support group of friends, who will remind you, that if you lose your job, you are not alone, and your long-term goals are important.
- Leverage your skills and your experiences so your successes can be replicated.
- Build a financial reserve for times when you may have to survive on your savings. Do not live on a day-to-day, hand-to-mouth financial basis.
- Acquire financial education to create personal financial goals and plan your future.

---

[2] https://doi.org/10.1190/tle40010010.1, Maria A. Capello, Anna Shaughnessy, and Emer Caslin, The Geophysical Sustainability Atlas: Mapping geophysics to the UN Sustainable Development Goals, *The Leading Edge*, Volume 40, Issue 1, January 2021, pp 1–80.

[3] https://en.wikipedia.org/wiki/French_Institute_of_Petroleum.

- We will not remain young forever. Begin planning for your retirement early in your career, even if it seems far away.
- Do not fear change. Resilience comes from being flexible and open to change and trust that you can navigate changes with confidence.

13.1 Add to the Bottom Line

During the expansionary, "boom time" phase of a business cycle, organizations can accumulate a lot of "fat". When there is plenty of money, managers may find it easier to ignore troublemakers and under-performing staff. At the beginning of a recession, those people are often the first to go and the impact on the survivors is minimal, because those terminated were non-productive "dead wood." The remaining workers are not demoralized, because many of them may have been wondering how the poor performers lasted as long as they did.

The longer and deeper the downturn, the more difficult the staff reductions become. When surviving staff see the termination of people who they consider to be strong contributors, they may ask themselves, "*Why them and not me?*".

Eve survived two rounds of layoffs in which half the people in her group were terminated. The remaining staff were demoralized and frightened. After the first of those two devastating layoffs, Eve asked to leave research and switch to a business function. The evening before the second fifty percent layoff, Eve spoke with a colleague who was still working in the job function Eve fled when the first 50% layoff occurred. He was confident that he would not lose his job, because they were short-staffed, and he was working overtime. What he didn't appreciate was that the company had decided to outsource that technical function. He was terminated and offered part-time work as a contractor.

When times are tough, organizations focus on their core strengths to survive. A tried-and-true method for individuals to survive is to "follow the money" and work on projects that are critical to the financial success of the organization.

13.2 Problem Solvers

As we mentioned in Chapter 2, the famous architect, I. M. Pei said, "*Success is a collection of problems solved.*"[4] If you can solve critical problems, you will always find work and it can be fascinating and exciting.

When Eve began her career, she thought she wanted to be a research scientist. With time, she felt she was learning more and more about less and less. Then she realized that if she took on technical service work, her work was just as interesting, and the business units would gladly increase funding if she was producing results that solved major problems. Eve realized that what she really enjoyed was making a large positive impact by solving serious problems. She ignored her narrow technical discipline and taught herself whatever was needed to solve the problem at hand and/or partnered with people whose skills complimented hers. In the next phase of her career, when she moved into management, she was eager to take on a wide range of assignments, because she knew she could quickly educate herself and become a problem solver. It helped her also, to have wonderful mentors along the way, and they helped enable her to realize her value amid criticism and other difficulties. On top of that, her experience made her a mentor, who asked younger professionals probing questions, to help them envision what kind of future they wanted for themselves, and what skills they already had to achieve their goals.

Maria Angela recognized she was a problem solver when she was given the opportunity to lead teams. Her strength lies in collaborative leadership. Although, she suffered low self-esteem and severe bullying in elementary school, she blossomed as an adult, leaving behind her fears and shyness, to energetically inspire and motivate teams. She enabled her teams to be enthusiastic about themselves and their objectives. When women needed empowerment, she created a network. When operators didn't trust new technologies, she launched a pilot project, so that all the stakeholders could ask questions and make modifications. Her insight grew with time, and she appreciates the guidance she received from multiple mentors over the course of her career. They helped her make difficult decisions about changing jobs and moving to another continent to advance her career. Mentors were pivotal for her, and her family was central to her analysis of alternatives.

---

4 Daum, Kevin, 18 Inspiring Quotes from I. M. Pei, April 26, 2017, https://www.inc.com/kevin-daum/18-inspiring-quotes-from-im-pei.html, as of December 15, 2021.

13.3 Lifelong Learning

Commit yourself to lifelong learning. The skills that are in demand at the beginning of your career may be under-valued and in over-supply later. You need to push yourself to constantly keep learning.

Don't be afraid to change directions, to take on unfamiliar work and to learn what it takes to do it well. If you have reinvented yourself previously, each successive reinvention becomes easier. It is amazing how seemingly unrelated experiences combine with each other to provide valuable insights.

Our lessons, especially at work, do not come from a classroom. They come from real-life feedback. Curiosity should drive our interest in technology developments and other innovations. We should nurture the desire to acquire new skills.

- Learn about new developments in your organization
- Practice self-education to continually update your skills
- Expand your knowledge beyond your area of expertise

13.4 Ombuds Programs

If you feel that you must bring a personal problem to the attention of someone in your organization and your organization has an Ombuds program, that can be an excellent source of assistance.

Some organizations including universities, have Ombuds programs. There is an enormous difference between the support you get from Human Resources and from an Ombuds program. The mission and purpose of Ombuds programs are drastically different from Human Resources groups.

The word ombudsman[5] comes from old Norse. King Charles XII of Sweden created an ombudsman program in 1713 to ensure that judges and civil servants acted in accordance with the laws and their duties. In the early 1800s, Norway and Sweden created parliamentary Ombuds programs to safeguard the rights of citizens by establishing a supervisory agency that was independent of the executive branch.

In short, Human Resources exists to serve management and Ombuds programs exist to aid staff members and in universities to students and

---

[5] Lang, C. McKenna, "A Western King and an Ancient Notion: Reflections on the Origins of Ombudsing," *Journal of Conflictology*, Volume 2, Issue 2 (2011) ISSN 2013–8857, November 2011 (http://openaccess.uoc.edu/webapps/o2/bitstream/10609/12627/1/06_Lang_ed.pdf).

staff. Merriam Webster dictionary defines an ombudsperson as *"a person who investigates, reports on, and helps settle complaints."*[6]

The Code of Ethics of the International Ombudsman Association[7] (Fig. 13.1) is shown below. If your organization's Ombuds program is functioning in alignment with good practices, it is supposed to be independent, neutral and impartial so that it can help resolve conflicts in a confidential and informal way. Formal grievance procedures managed by Human Resources tend to be weighted in favor of management.

If your organization's Ombuds program complies with the International Ombudsman Association Code of Ethics which specifies holding in complete confidence all communications with those who seek help and not disclosing those confidential discussions unless expressly given permission to do so, it can be a very valuable benefit. A well run Ombuds program provides access to advice from a person who understands the politics of your organization. The ombudsperson can provide suggestions on how to manage the conflict and whether you should file a formal grievance. You often have more alternatives than you realize. Sometimes, even though you are clearly the injured party, it is better to "move on" than to fight.

Eve spent thirty-five years working for giant corporations. Early in her career, she didn't know better and sought advice from Human Resources (HR). She didn't have access to an Ombuds program until near the end of her career. The problem she took to the Ombudswoman was handled much better than the earlier issues. The Ombudswoman suggested serving as an intermediary and expressly sought Eve's permission about everything she discussed with the company.

Effectively functioning whistleblower[8,9] and ombudsman policies are ways of protecting the rights of employees and a positive step towards ensuring healthy work environments. Modern organizations that seek to be resilient, should protect their workforce with robust ombuds and whistleblower programs.

The cultural aspects of the countries or regions where the program will be implemented, must be taken into consideration. In some locations including the Middle East and Latin America, there is skepticism about the confidentiality of (denouncing) whistleblower processes. These societies are highly interconnected and networked. Almost everyone knows everybody

---

[6] https://www.merriam-webster.com/dictionary/ombudsperson, as if December 13, 2021.

[7] https://www.ombudsassociation.org/assets/IOA%20Code%20of%20Ethics.pdf as of December 13, 2021.

[8] https://www.merriam-webster.com/dictionary/whistleblower.

[9] https://corporatefinanceinstitute.com/resources/knowledge/other/whistleblower-policy/.

INTERNATIONAL OMBUDSMAN ASSOCIATION

IOA CODE OF ETHICS

PREAMBLE

The IOA is dedicated to excellence in the practice of Ombudsman work. The IOA Code of Ethics provides a common set of professional ethical principles to which members adhere in their organizational Ombudsman practice.

Based on the traditions and values of Ombudsman practice, the Code of Ethics reflects a commitment to promote ethical conduct in the performance of the Ombudsman role and to maintain the integrity of the Ombudsman profession.

The Ombudsman shall be truthful and act with integrity, shall foster respect for all members of the organization he or she serves, and shall promote procedural fairness in the content and administration of those organizations' practices, processes, and policies.

ETHICAL PRINCIPLES

INDEPENDENCE

The Ombudsman is independent in structure, function, and appearance to the highest degree possible within the organization.

NEUTRALITY AND IMPARTIALITY

The Ombudsman, as a designated neutral, remains unaligned and impartial. The Ombudsman does not engage in any situation which could create a conflict of interest.

CONFIDENTIALITY

The Ombudsman holds all communications with those seeking assistance in strict confidence, and does not disclose confidential communications unless given permission to do so. The only exception to this privilege of confidentiality is where there appears to be imminent risk of serious harm.

INFORMALITY

The Ombudsman, as an informal resource, does not participate in any formal adjudicative or administrative procedure related to concerns brought to his/her attention.

www.ombudsassociation.org

Fig. 13.1 Code of Ethics of the IOA

else, Secrets at work may easily be disclosed in social settings and through myriad channels reach the ears of the offenders at work. Nevertheless, the risk is worthwhile because silence is never the answer to these issues.

14

You Have More Alternatives Than You Think

Years ago, an executive who Eve knew slightly lost his job and committed suicide. Eve had the impression he was very successful and was shocked that he killed himself. Apparently, his attempts to get an equivalent position to one he lost were unsuccessful and he couldn't imagine an alternative path to success.

When your hunt for a new job is fruitless, it is easy to become depressed. Don't get trapped in a downward spiral. Where there is life, there's hope. Don't limit your search to positions similar to what you have been doing. The broader your job search, the greater your chances of success.

Look at the situation in a positive way. It is an opportunity to reinvent yourself. Are there things you always dreamed about doing, but couldn't in your previous position? Think about how you can include those in a new career. You can be more successful in pursuing your dreams than settling for a position that you feel is a demotion.

Maria Angela learned in her many changes of jobs and countries that success is and should be intrinsic to the person. Success belongs to the individual and is a product of her or his accomplishments. You can be very successful in life, even if you do not have an exalted title with an employer. She looks back at her career and has asked herself the question, "What if I had attached my concept of success to becoming a director in PDVSA? Or even to growing professionally in Venezuela?" As a professional who had to emigrate and find new opportunities because she could not continue to work in her own country, she would have felt devastated. Her life journey

E. Sprunt and M. A. Capello, *A Guide to Career Resilience*, https://doi.org/10.1007/978-3-031-05588-1_14

followed a path she never anticipated. Her confidence that she could reinvent herself increased as she continued to build her career, overcoming challenges that sometimes were predictable and at times were beyond her own wildest scenarios.

14.1 What Do You Really Want to Do?

Even if you have not lost or are not in danger of losing your job, it is beneficial periodically to think about whether you want to investigate alternative positions. Over time, our priorities and interests change. Even if you are not in the middle of an employment crisis, periodically exploring your alternatives can be beneficial. Ask yourself, not only which skills are your strengths but also what do you enjoy doing and what have you always hoped to accomplish?

The type of work you thought you wanted at the beginning of your career may no longer be satisfying. Ask yourself:

- What do I like about my current job?
- What do I dislike about my current job?
- What has changed in my personal life since I accepted this job?
- Is the current and/or emerging leadership of my organization aligned with my ethical principles?
- Do I perceive my current company to be resilient?
- What would I like to do, that I can't in my current role?

Early in her career, Eve wanted to do research, but gradually realized that it could be much more exciting to do technical service. Persuading a business unit to try new technology if they have a proven existing solution can be extremely difficult. They'd rather "stay with the devil they know" than take a risk on new technology. In contrast, when a business unit asked Eve to help them with a problem to which they had no satisfactory solution, they were eager to adopt the new technology she proposed. Eve realized that what she enjoyed was being a problem solver. She wasn't focused on pursuing a narrow research specialty.

Maria Angela has had to reinvent herself multiple times. Sometimes, unexpectedly and because of personal issues or her local country environment, she has faced tough decisions. She began in the exploration department of the biggest operator in Venezuela, but for personal reasons, switched to the research center of the same company where she became interested in reservoir optimization loops and retrained herself from exploration to production.

As a result of the political situation in Venezuela, she reinvented herself several more times. Recently, in what she considers her most interesting re-invention, she began focusing on sustainability, following international trends and personal ethical goals. She enlarged her network, joined committees related to sustainability topics, and has gotten certified in sustainability. Change is invigorating, and many times, as Maria Angela discovered, the best way to refresh your career.

14.2 Assess Your Strengths

In preparation for a career change, ask yourself what the major factors were in your previous career successes. Sometimes they are not what you might have assumed. Identifying and developing a description of character attributes that have enabled you to be successful in the past prepares you for job interviews for new types of work. Many of your talents are transferrable to a wide range of applications and can be of tremendous value in a dramatically different career.

When Eve looked back at what she did that had the greatest impact, there were common patterns. She realized she was an open-minded observer and was willing to challenge the applicability of previous assumptions. About five years into her career, she was responsible for supervising routine tests to be done by a commercial laboratory. When she examined the samples, she realized standard measurement techniques done on them would produce misleading results. Her predecessors had noted the problems, but "blindly" followed orders and applied standard procedures that Eve showed yielded misleading results. Eve raised an alert that had a major positive financial impact on her company.

Eve was trained to be a research earth scientist but realized she could apply her research training to anything. When she was diagnosed with a rare metastatic cancer, she turned her ability to read technical papers to reading medical journals and scrutinizing her case. She took an active role in decisions about what types of treatment would be best. With her knowledge of laboratory quality control issues, she didn't panic when a giant commercial medical laboratory produced a series of measurements indicating that she had a new rapidly growing tumor. She had the tests repeated in another lab and demonstrated that the first lab's results were all erroneous.

At work, Eve uncovered numerous previously ignored errors in earth science data that had led to misinterpretation of what was happening in the subsurface. While struggling with a rare metastatic cancer, Eve discovered critical factors that had been overlooked by her doctors. Her husband helped her gather all the records from her surgeries and hospitalizations. Their diligence paid off. When she was discharged from the hospital after a month, the discharge nurse didn't tell her to follow-up on what turned out to be a serious pulmonary metastasis. Reading all her diagnostic test results, Eve found the notation recommending additional tests on her lungs. Eve believes her success in beating her grim prognosis could be attributed to applying the same research skills that led to her greatest successes at work. Eve collected all the available data on her condition, carefully read all her test results, requested modifications to her treatment, and spoke-up when she observed errors and dangerous practices. Her experience as a mentor also helped her in guiding the conversations with her doctors about her preferred choice of treatments.

Over the course of her career, Eve often assumed roles previously held by someone with a different technical background. Whether it was a technical problem, a business issue, a public relations challenge or recruiting, Eve independently reviewed the data and reassessed the strategy. She did not blindly follow orders. She questioned assumptions, quality checked, and reassessed data. That approach didn't make her life easy. Challenging previous assumptions and questioning decisions does not always win friends. At work, she learned to focus her efforts on the errors that had a major financial impact on her employer.

You will benefit from assessing your strengths and a mentor can help with that. Often other people observe strengths that you fail to recognize in yourself. Prioritize your concerns about your career development in relation to your strengths. This mapping can be shared with your mentor, who can help you assess your strengths from another perspective. Our life experiences are often useful, even for paths that may be dramatically different from our academic origins.

Maria Angela is a physicist, and her master's degree was in geophysics, but she is recognized as an expert in reservoir management, diversity and inclusion, empowerment of women, and sustainability. Her career has taken many unexpected turns, and she has grasped opportunities to grow in different directions depending on what could benefit her work and her employer.

14.3 Reinventing Yourself

If you do an inventory of your skills, you will find that many of them are transferrable to different roles. Instead of plodding along, stuck in a rut, rejuvenate yourself by taking a risk to follow a new opportunity.

On Eve's twentieth anniversary with Mobil Oil Corporation, the company announced its intent to merge with Exxon. Years earlier, Mobil had implemented a secret "poison pill" to discourage an outsider from acquiring the company. If Mobil was acquired, any employee who was fired, demoted or moved more than fifty miles could get two weeks' pay per year of service. Eve seized the opportunity to reinvent herself. It was one of the best decisions she ever made.

15

No Role Model, No Problem

Not having role models can be extremely liberating, because if you seize an opportunity, it is up to someone else to take the risk to stop you and many people are risk averse.

15.1 Personal Priorities

For much of her career, Eve had no women in front of her to mimic, so she had to forge her own path. If you know your priorities, it is easier to decide which risks you wish to take. Eve placed a very high priority on getting her doctorate., but as soon as she accomplished becoming the first woman to get a Ph.D. in geophysics from Stanford, she moved on to the next thing on her list, which was having a baby while continuing her career progression and living in the same geographic area as her husband. Her husband had a couple more years to go to finish his MBA and law degrees at Stanford, so after she completed her Ph.D., Eve arranged to continue as a research associate at Stanford. She didn't share that she planned to become pregnant and hid her pregnancy as long as possible.

Her son, Alexander, was born nine months after she defended her Ph.D. thesis. When he was thirteen days old, Eve brought him to work with her in a front baby pack. She continued to bring her baby to work with her every day until he was about six months old and very active. Then she found a babysitter.

© The Author(s), under exclusive license to Springer Nature Switzerland AG 2022
E. Sprunt and M. A. Capello, *A Guide to Career Resilience*,
https://doi.org/10.1007/978-3-031-05588-1_15

No one would have dared to officially grant her permission to bring a baby into a building filled with laboratories. Eve didn't want to stop her scientific work, and she didn't want to forgo precious time with her infant, so she didn't ask anyone to approve her actions. What surprised her was that when she stopped bringing Alexander to work, many people complained they missed him. Eve had worried Alexander would cry in the library, but the librarians complained when they no longer got to see him!

15.2 Be Willing to Take Risks

With no role models to follow, Eve thought it would be easier to combine motherhood with a career in industry than motherhood and an academic career because of the time limit to get tenure. While working on her Ph.D. thesis, Eve regularly met with interviewers. One day during interview season, she saw a fellow female student in tears. The interviewer had asked her, "What's a nice girl like you doing in geophysics?" Eve always wondered why she was never asked that question. Instead, the interviewers always illegally probed her about her maternity plans. So, for her first job interview after Alexander's birth, she selected a time when he was most likely to be asleep and wore him in the infant pack to the interview. She told the interviewer, "*Here's the question you are not supposed to ask!*" The interviewer's reaction was perfect. He invited the entire family for the interview trip and arranged babysitting. For subsequent interview trips, Eve asked the company she was visiting to provide babysitting for Alexander, who always accompanied her, and they did so at no charge to her.

15.3 Client-Pleasing Performance

Maria Angela is known for acting many times without asking permission. This got her in serious trouble while she was working in INTEVEP, the research center of PDVSA in Venezuela. While working at INTEVEP, Maria Angela mimicked the way service companies presented reprocessed seismic data. She hung up big posters of the "pre" and "post" processed data in a conference room, scheduled breaks with coffee and snacks, and invited a director to open the meeting. No one else had done something like that before. The cost of the banners, the coffee and the catering had prevented other people from organizing similar presentations, because they never thought of acting like a private company, but Maria Angela did.

Maria Angela didn't ask for her department head's permission or approval of the catering and the printed posters before placing the orders. When people in the service departments asked for signatures, she told them she was inviting the Director and would have him sign all the catering forms at the presentation. It was certainly unorthodox, but it worked. She implemented a new style of approaching the clients without permission and got an extension and expansion of the contract for Intevep by raising her presentation to a higher standard with a better marketing strategy. Of course, she had to apologize to her direct supervisor, but it was with the acknowledgement that the strategy had won a new contract for her department, so her boss was pleased to do so.

15.4 Investment of Personal Time

Even if you plan to do something that is related to your job on your personal time, companies often expect you to request permission in advance. While Maria Angela was working for Kuwait Oil Company, she agreed to serve as the chair of several committees without advance permission. At KOC, this type of appointment usually required permission from a long chain of people. She decided that because she was going to use her personal time for the projects and was not representing the company, she would not seek official approval. She leveraged the time difference between the USA and Kuwait and attended meetings using remote videoconferencing long before the use of Zoom and Teams became widespread during the pandemic.

To avoid requesting advance permission to attend more conferences than her management was likely to approve, Maria Angela used her vacation time to attend and took advantage of the reduced registration fees for speakers by always presenting a paper. Her family learned that vacations would usually be in conjunction with the Society of Petroleum Engineers (SPE), the Society of Exploration Geophysicists (SEG) or the American Association of Petroleum Geologists (AAPG) annual meetings. She never requested formal permission to serve as the chair of any of the numerous professional society committees she led. If she had sought permission, it probably would not have been granted, because she took on so many roles.

15.5   No Role Models, No Problem

Dictionary.com defines a role model as, "a person whose behavior, example, or success is or can be emulated by others, especially by younger people."[1] When as a pioneer, you have no role models, you must allow yourself the freedom to do what you think is best and in forging a new path, you become a role model for those who follow you.

When you are breaking new ground without role models, you must give yourself permission to take risks. For people in under-represented groups in certain work environments, if you have courage, the lack of role models can liberate you to take the risks that you believe are necessary to succeed.

Eve was not the first woman to attempt to get a Ph.D. in Geophysics from Stanford, but she was the first to succeed. When you are the first person with your characteristics to take on challenges, you must be aware that you have no role models and will have to figure out what works for you. Simply mimicking the behavior of typical representative of the majority may not work at all for members of under-represented minorities.

Often during their careers, both Eve and Maria Angela were the first woman doing things that no woman had done before. They had to decide for themselves how to achieve the results they thought were necessary to advance their careers.

15.6   No Boundaries

Geophysics, which both Maria Angela and Eve studied is in its very name a multidisciplinary field. Whatever your diploma lists as your major, it doesn't mean that is what you necessarily are destined to be for the rest of your life. The moment you graduate, your knowledge begins to go stale unless you continually reeducate yourself. With a good education, you can teach yourself almost anything.

Don't feel trapped by your choice of major in college. Technology is changing so rapidly we must constantly be learning the new technology and skills needed to be competitive at work. Follow your dreams and teach yourself what you need to know and, if necessary, go back for extra training to get any certifications that you need.

In large organizations, it is often possible to get a transfer into a role for which you don't meet all the requirements. Go ahead, apply for positions

---

[1] https://www.dictionary.com/browse/role-model as of December 3, 2021.

of interest to you for which you do not currently have all the qualifications. Women tend to worry more than men about whether they meet all the qualifications listed for a position. If you feel stuck in a rut, don't wait for permission to apply for a job of interest to you. Give yourself permission to apply and make a commitment to yourself to put in the effort to learn what you need to excel in the new role. Learning new things is very rejuvenating and energizing.

Furthermore, it is often amazing how having a diverse background provides insights and ideas for new approaches to solving problems that do not occur to people with more limited experience within a narrow specialty. Broad knowledge is often extremely beneficial in devising solutions to difficult problems, because you see beyond a narrow range of factors and methodologies.

15.7 Be Fearless

When you are a trailblazing representative of a minority, you have a lot more to worry about than the typical person who fills a role. The typical employee can focus on just doing a good job. Minority members not only have to excel in their work, they also must shape the expectations of those around them.

Take responsibility for your actions. Do not bother supervisors, by asking for permission for small concessions that make it possible for you to do your job. If you aspire to rise to positions of authority, recognize that involves taking responsibility for your actions according to your beliefs, ethical grounds, and morals. The first step to becoming a leader, is to behave like one. But be ready to explain to your supervisor what you are doing and why.

16

Rules to Live By

Here are some of the rules that Maria Angela and Eve learned working in a male-dominated industry in which they were pioneers.

16.1 General Principles

- We are not born resilient. We become resilient.
- Focusing on long-term goals makes us more resilient, because short-term issues are minimized in contrast with our long-term objectives.
- Incorporate stress reduction routines into your daily life, such as exercise, music, hiking and other activities that can help you cope with stress and enhance your physical strength to face adversity.
- Understand your strengths and weaknesses, so you can leverage your strengths, and resiliently respond to tough times.
- Build a support group who will remind you that if you lose your job, you are not alone, and your long-term goals are important.
- Leverage your skills and your experiences so your successes can be replicated.
- Build a financial reserve for times when you may have to survive on your savings.

- Acquire financial education to create personal financial goals and plan your future.
- Do not live on a day-to-day, hand-to-mouth financial basis.
- We will not remain young forever. Begin planning for your retirement early in your career, even if it seems far away.
- Do not fear change. Resilience comes from being flexible and open to change—you can navigate changes with confidence.

16.2 Lifelong Learning

Commit yourself to lifelong learning. The skills that are in demand at the beginning of your career may be under-valued and in over-supply later. You need to push yourself to keep learning new skills.

Don't be afraid to take on unfamiliar work and learn what it takes to do it well. If you have reinvented yourself previously, each successive reinvention becomes easier. It is amazing how seemingly unrelated experiences combine with each other to provide valuable insights.

Our lessons, especially at work, do not come from a classroom. Curiosity should drive our interest in technology developments and other innovations. We should nurture the desire to acquire new skills.

- Learn about new developments in your organization
- Practice self-education to continually update your skills
- Expand your knowledge beyond your area of expertise

16.3 Networking

Your network is your safety net. No matter how well you are doing within your organization, be sure to build a network that extends far beyond its borders. Consider building and maintaining your network to be an essential part of your career plan.

Networks that are not used decay. A healthy network requires ongoing maintenance. Contact key people regularly. When they call you for help, do whatever you can even if it you are incredibly busy. Do favors never expecting one in return. Pay it forward and hope that you never need help yourself.

16.4 Magic Three

Your keys to lifelong success are:

- Lifelong learning
- Problem solving
- Networking.

If you pay attention and develop all three, you will remain valuable and land on your feet whenever you are hit by a storm. You will be resilient!

In contrast with other dramatic shifts in her life and career, that were triggered in reaction to political, health and commuting issues, this time, Maria Angela planned her move leveraging her rules on how to become resilient.:

- We are not born resilient. We become resilient.
- Focusing on long-term goals makes us more resilient, because short-term issues are minimized in contrast with our long-term objectives.
- Incorporate stress reduction routines in your daily life. Exercise, especially outdoor exercise such as walking and hiking, and hobbies likes playing a musical instrument or listening to music can help you cope with stress and enhance your physical strength to face adversity. Find something that you like to do on a regular basis and make carving out time to do it a priority.
- Understand your strengths and weaknesses, so you can leverage your strengths, and resiliently respond to tough times.
- Build a support group who will remind you if you lose your job that you are not alone, and that your long-term goals are important.
- Leverage your skills and your experiences so your successes can be replicated.
- Build a financial reserve for times when you may have to survive on your savings.
- Acquire financial education to create personal financial goals and plan your future.
- Do not live on a day-to-day, hand-to-mouth financial basis.
- We will not remain young forever. Begin planning for your retirement early in your career, even if it seems far away.
- Do not fear change. Resilience comes from being flexible and open to change - you can navigate changes with confidence.
- Learn about new developments in your organization
- Practice self-education to continually update your skills
- Expand your knowledge beyond your area of expertise

16.5 Your Health

When you are young, you can abuse your body and get away with it for a while. As you get older, it takes longer for your body to recover from lack of sleep, lack of exercise and poor nutrition. No matter how busy you are, take the time necessary to exercise and maintain your physical condition. Exercise and adequate sleep can help you think more clearly and have the stamina to solve difficult problems.

Don't ignore physical problems because you don't have time to deal with them. If you catch serious illnesses including aggressive, life-threatening cancers early, you have a much better chance of surviving and regaining your quality of life. Delaying seeking medical treatment to meet work-related deadlines can be the difference between life and death. Modern medicine can work miracles, but you must do you part by seeking medical advice when you notice changes in your body.